Ramona Jochmann

O-linked N-Acetylglucosaminylation of Sp1 Inhibits the HIV-1 Promoter

AF062216

Ramona Jochmann

O-linked N-Acetylglucosaminylation of Sp1 Inhibits the HIV-1 Promoter

Sweet Perspectives to Repress the Human Immunodeficiency Virus Type-1

Südwestdeutscher Verlag für Hochschulschriften

Impressum / Imprint
Bibliografische Information der Deutschen Nationalbibliothek: Die Deutsche Nationalbibliothek verzeichnet diese Publikation in der Deutschen Nationalbibliografie; detaillierte bibliografische Daten sind im Internet über http://dnb.d-nb.de abrufbar.
Alle in diesem Buch genannten Marken und Produktnamen unterliegen warenzeichen-, marken- oder patentrechtlichem Schutz bzw. sind Warenzeichen oder eingetragene Warenzeichen der jeweiligen Inhaber. Die Wiedergabe von Marken, Produktnamen, Gebrauchsnamen, Handelsnamen, Warenbezeichnungen u.s.w. in diesem Werk berechtigt auch ohne besondere Kennzeichnung nicht zu der Annahme, dass solche Namen im Sinne der Warenzeichen- und Markenschutzgesetzgebung als frei zu betrachten wären und daher von jedermann benutzt werden dürften.

Bibliographic information published by the Deutsche Nationalbibliothek: The Deutsche Nationalbibliothek lists this publication in the Deutsche Nationalbibliografie; detailed bibliographic data are available in the Internet at http://dnb.d-nb.de.
Any brand names and product names mentioned in this book are subject to trademark, brand or patent protection and are trademarks or registered trademarks of their respective holders. The use of brand names, product names, common names, trade names, product descriptions etc. even without a particular marking in this work is in no way to be construed to mean that such names may be regarded as unrestricted in respect of trademark and brand protection legislation and could thus be used by anyone.

Verlag / Publisher:
Südwestdeutscher Verlag für Hochschulschriften
ist ein Imprint der / is a trademark of
OmniScriptum GmbH & Co. KG
Heinrich-Böcking-Str. 6-8, 66121 Saarbrücken, Deutschland / Germany
Email: info@svh-verlag.de

Herstellung: siehe letzte Seite /
Printed at: see last page
ISBN: 978-3-8381-1144-5

Zugl. / Approved by: Erlangen, Friedrich-Alexander Universität Erlangen-Nürnberg, Dissertation, 2009

Copyright © 2009 OmniScriptum GmbH & Co. KG
Alle Rechte vorbehalten. / All rights reserved. Saarbrücken 2009

Table of Contents

1. SUMMARY 1
2. INTRODUCTION 3
   2.1 The Human Immunodeficiency Virus Type-1 3
      2.1.1 The Structure of HIV-1 3
      2.1.2 The HIV-1 Proteins 4
      2.1.3 The Regulation of HIV-1 Gene Expression 5
      2.1.4 The HIV-1 Replication Cycle 6
      2.1.5 The Clinical Course of HIV-1 Infection 6
      2.1.6 The Highly Active Antiretroviral Therapy 7
      2.1.7 Residual Viremia and HIV-1 Latency 9
   2.2 The Transcription Factor Sp1 13
      2.2.1 Characteristics of Sp1 13
      2.2.2 Transcriptional Activation by Sp1 15
      2.2.3 Regulation of the Transcriptional Activity of Sp1 16
         2.2.3.1 Regulation of Sp1 Expression Level 17
         2.2.3.2 Regulation by Posttranslational Modifications 17
         2.2.3.3 Regulation by Interaction Partners 19
   2.3 Posttranslational Modification of Proteins with O-GlcNAc 20
      2.3.1 The Hexosamine Biosynthesis Pathway 20
      2.3.2 The Cycling Enzymes Involved in O-GlcNAc Modification of Proteins 22
         2.3.2.1 O-GlcNAc Transferase 22
         2.3.2.2 O-GlcNAc Hexosaminidase 23
         2.3.2.3 Interplay Between NCOAT and OGT at the Chromatin 24
      2.3.3 Diverse Regulation of Protein Function by O-GlcNAc 25
      2.3.4 The Yin-Yang Hypothesis 28
   2.4 Aims of the Study 30
3. MATERIAL AND METHODS 31
   3.1 Material 31
      3.1.1 Chemicals, Media and Reagents 31
      3.1.2 Consumables 33
      3.1.3 Equipment 33
      3.1.4 Kits 34
      3.1.5 Enzymes 34
      3.1.6 Standards 35
      3.1.7 Software 35
      3.1.8 Buffer and Solutions 35
      3.1.9 Primary Antibodies 37
      3.1.10 Secondary Antibodies 37
      3.1.11 Oligonucleotides 38
         3.1.11.1 siRNA-Oligonucleotides 38
         3.1.11.2 Cloning Oligonucleotides 38
         3.1.11.3 Gel Shift Oligonucleotides 38
         3.1.11.4 Oligonucleotides for RT-PCR 39
         3.1.11.5 Sequencing Oligonucleotides 39
      3.1.12 Plasmids 40
      3.1.13 Biological Material 43

- 3.1.13.1 Bacterial Strains ... 43
- 3.1.13.2 Eukaryontic Cell Lines ... 43
- 3.1.13.3 Primary Cells ... 44
- 3.1.13.4 Recombinant Viruses ... 44

3.2 Methods ... 44
3.2.1 RNA-Techniques ... 44
- 3.2.1.1 Isolation of Cellular RNA from Eukaryotic Cells ... 44
- 3.2.1.2 Agarose Gel Electrophoresis of RNA ... 44
- 3.2.1.3 Quantitative Measurement of RNA ... 45
- 3.2.1.4 In vitro Transcription of mRNA ... 45

3.2.2 cDNA Synthesis ... 46
3.2.3 DNA-Techniques ... 46
- 3.2.3.1 Polymerase Chain Reaction (PCR) ... 46
    - a) Reverse Transcription (RT)-PCR ... 46
    - b) Cloning-PCR ... 47
    - c) Mutagenesis-PCR ... 48
- 3.2.3.2 Agarose Gel Electrophoresis of DNA ... 49
- 3.2.3.3 Isolation of DNA Fragments from Gels ... 49
- 3.2.3.4 Restriction Digest of DNA Molecules ... 49
- 3.2.3.5 Purification of Linearized DNA with Phenol/Chloroform/Ethanol ... 50
- 3.2.3.6 Dephosphorylation of Vector DNA ... 51
- 3.2.3.7 Klenow Reaction ... 51
- 3.2.3.8 Ligation of DNA ... 51
- 3.2.3.9 Heat Shock Transformation of Bacteria ... 51
- 3.2.3.10 Isolation of Plasmid DNA ... 52
- 3.2.3.11 DNA Sequencing ... 52
- 3.2.3.12 Quantitative Measurement of DNA ... 53

3.2.4 Cell Biological Methods ... 53
- 3.2.4.1 Cultivation of E.Coli ... 53
- 3.2.4.2 Cultivation of Mammalian Cells ... 53
- 3.2.4.3 HIV-1 Pseudovirus Production ... 54
- 3.2.4.4 Infection of Lymphocytes with HIV-1 and Stimulation with GlcN ... 54
- 3.2.4.5 Cell Transfection ... 54
    - a) Transfection of HEK 293T Cells ... 54
    - b) Transfection of Jurkat Cells ... 55
    - c) Transfection of $CD4^+$ Primary T Lymphocytes with mRNA ... 55

3.2.5 Protein Chemistry ... 56
- 3.2.5.1 Determination of Protein Concentration ... 56
- 3.2.5.2 SDS-Polyacrylamide Gel Electrophoresis (SDS-PAGE) ... 56
- 3.2.5.3 Western Blot Analyses ... 56
- 3.2.5.5 Luciferase Reporter Gene Assay ... 57
- 3.2.5.6 Flow Cytometry Analyses ... 57
- 3.2.5.7 Cellular Analysis ... 58
    - a) CytoTox 96 Non-Radioactive Cytotoxicity Assay ... 58
    - b) CellTiter 96 Non-Radioactive Cell Proliferation Assay ... 58
- 3.2.5.9 Nuclear/cytosol Fractionation ... 59
- 3.2.5.10 Immunoprecipitation ... 59
- 3.2.5.11 Electrophoretic Mobility Shift Assay (EMSA) ... 59
    - a) Hybridization of Oligos ... 59
    - b) Labeling of Oligos ... 59
    - c) Band Shift Assay ... 60

3.2.6 Statistical Analyses ... 61

4. RESULTS 62
   4.1 GlcN Inhibits HIV-1 Transcription in Lymphocytes *62*
   4.2 GlcN Represses HIV-1 LTR Promoter Activity in HeLa Cells *64*
   4.3 OGT Inhibits HIV-1 Transcription in Primary Lymphocytes *65*
   4.4 The Sp1-Binding Sites Are Necessary and Sufficient for the Inhibition of HIV-1 LTR Activity by OGT *66*
   4.5 Sp1 Is Necessary for the Inhibitory Effect of OGT on the HIV-1 LTR *70*
   4.6 O-GlcNAcylation of Sp1 Selectively and Dose-Dependently Inhibits the HIV-1 LTR *71*
   4.7 OGT Does Not Inhibit Expression and DNA-Binding Affinity of Sp1 .... *72*
   4.8 Sp1-O-GlcNAcylation Is Necessary for the Inhibition of the HIV-1 LTR *74*
   4.9 Increased OGT Expression Before Infection Increases OGT-Inhibitory Effect on HIV-1 Replication *77*
5. DISCUSSION 79
   5.1 O-GlcNAcylation Inhibits HIV-1 Transcription *79*
   5.2 Sp1-Binding Sites and Sp1 Protein Are Necessary for the O-GlcNAc-Mediated Inhibition of HIV-1 Transcription *80*
   5.3 O-GlcNAcylation of Sp1 Inhibits the HIV-1 LTR in a Dominant Manner *81*
   5.4 O-GlcNAcylation of Sp1 Specifically Inhibits HIV-1 LTR Activity *82*
   5.5 Mechanism for O-GlcNAc-Mediated Inhibition of HIV-1 Transcription *83*
   5.6 Role of Sp1-O-GlcNAcylation in the Immune System in the Context of Viral Infection *85*
   5.7 Possible Implications of Sp1-O-GlcNAcylation in the Antiretroviral Therapy *86*
6. REFERENCES 88
7. ABBREVIATIONS 103
8. PUBLICATIONS 106
9. ACKNOWLEDGEMENTS 107

1. Summary

Human immunodeficiency virus-1 (HIV-1) gene expression and replication is regulated by the promoter/enhancer located in the U3 region of the proviral 5'-long terminal repeat (LTR). The binding of cellular transcription factors to specific regulatory sites in the 5'-LTR is a key event in the replication cycle of HIV-1. Since transcriptional activity is regulated by the posttranslational modification of transcription factors with the monosaccharide O-linked N-acetyl-D-glucosamine (O-GlcNAc), this study aimed to evaluate whether increased O-GlcNAcylation affects HIV-1 transcription.

In the present study it is demonstrated that treatment of HIV-1-infected lymphocytes with the O-GlcNAcylation-enhancing agent glucosamine (GlcN) repressed viral transcription in a dose-dependent manner. Overexpression of O-GlcNAc transferase (OGT), the sole known enzyme catalyzing the addition of O-GlcNAc to proteins, specifically inhibited the activity of the HIV-1 LTR promoter in different T cell lines and in primary $CD4^+$ T lymphocytes. Inhibition of HIV-1 LTR activity in infected T cells was most efficient (>95%) when OGT was recombinantly overexpressed prior to infection. O-GlcNAcylation of the transcription factor Sp1 and the presence of Sp1-binding sites in the LTR were found to be crucial for this inhibitory effect.

This study led to the conclusion that O-GlcNAcylation of Sp1 inhibits the activity of the HIV-1 LTR promoter. Modulation of Sp1 O-GlcNAcylation may play a role in the regulation of HIV-1 latency and activation, and links viral replication to the glucose metabolism of the host cell. Hence, the establishment of a metabolic treatment might supplement the repertoire of antiretroviral therapies against AIDS.

2. Introduction

2.1 The Human Immunodeficiency Virus Type-1

The human immunodeficiency virus type-1 (HIV-1) belongs to the Retrovirus family and is a member of the genus Lentivirus. Lentivirus infections have been described in several vertebrate species, where they cause immunodeficiency. HIV-1 and -2 are the two Lentivirus representatives infecting humans. They are envelope coated single-stranded RNA (ssRNA) viruses with double-stranded DNA (dsDNA) as intermediate product. Both were initially transmitted from monkeys to humans [HIV-1 is thought to have originated in southern Cameroon after jumping from wild chimpanzees to humans during the twentieth century (61, 117); HIV-2 may have originated from sooty mangebey (187)], where they evolved to human pathogens. However, HIV-2 has a lower virulence and transmittability than HIV-1.

The HIV-1 epidemic began in the 1980's in central Africa and the USA. It was recognized for the first time in 1981 as a new acquired cellular immunodeficiency in male homosexuals with opportunistic infections (71). HIV-1 was isolated in 1983 from a Caucasian patient with signs and symptoms that often precede the acquired immune deficiency syndrome (AIDS) (8). As yet, HIV-1 infection has spread worldwide, especially in developing countries, and reached pandemic levels: about 33 million people were infected with HIV by the end of 2007 (230).

2.1.1 The Structure of HIV-1

HIV-1 virus particles are roughly spherical with a diameter of about 120 nm (Fig. 1). The conical capsid is composed of the viral protein p24 and contains two copies of the viral genome, each with a size of 9 kb. The ssRNA copies are associated with the nucleocapsid proteins and enzymes needed for the development of the virion, such as reverse transcriptase (RT), protease, ribonuclease and integrase. A matrix composed of the viral matrix protein surrounds the capsid ensuring the integrity of the virion particle. The viral envelope is composed of phospholipids derived from the host cell membrane and is interspersed with proteins from the host cell and the viral envelope protein (Env). Three gp120 glycoprotein molecules form the cap of the Env protein, while the stem consists of three gp41 glycoproteins, which anchor the structure into the viral envelope (21). HIV-1 attaches *via* the Env protein to the CD4 receptor on the target cell, using the chemokine receptors CXCR4 (41) and CCR5 (56) as coreceptors.

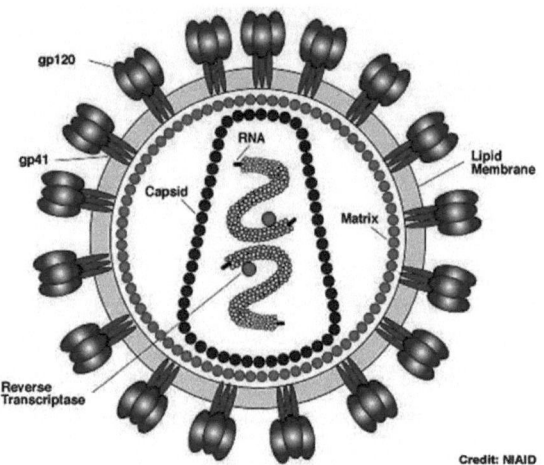

Fig. 1. Two dimensional structure of the HIV-1 virus particle. The conical capsid inside the virus particle contains two viral RNA-genomes, which are associated with nucleocapsid proteins and one molecule reverse transcriptase. The capsid is surrounded by an envelope containing the viral Env protein, which is composed of trimers of the glycoproteins gp120 and gp41. The inner surface is formed by matrix proteins. Figure credit by the National Institute of Allergy and Infectious Disease (NIAID).

2.1.2 The HIV-1 Proteins

HIV-1 encodes 15 proteins, grouped according to their function: (i) the structural proteins necessary for the viral assembly are synthesized as a precursor protein (group specific antigen, Gag), which is cleaved by the viral protease into the structural matrix-, capsid-, nucleocapsid- and linker-proteins. Similarly, (ii) the enzymes necessary for viral replication, the reverse transcriptase, the integrase and the protease are also produced by a precursor protein (polymerase, Pol), which is processed by the viral protease in the matured viral particles. (iii) The envelope (Env) precursor protein is cleaved by a cellular protease at the Golgi apparatus to the glycoproteins gp120 and gp41.

In addition, HIV-1 encodes several accessory proteins, of which two are essential: The transactivator of transcription (Tat) and the regulator of expression of viral proteins (Rev) enhance the transcription and elongation from the viral promoter and the shuttling of viral RNA into the cytoplasm, respectively (115, 133, 153). Four further accessory proteins, though not essential for the viral replication, increase viral infectivity or help the virus to escape the host's immune response. For example, the Vif protein prevents the action of APOBEC3G (154), a cellular protein which deaminates DNA:RNA hybrids. The Nef protein downregulates the CD4 receptor in order to facilitate the viral release (14), as well as the MHC class I and class II molecules (72, 218), thereby

Introduction

enabling escape from immune surveillance. The Vpr protein arrests cell division at G2/M (52). This is supposed to favor high expression of viral genes. And finally, Vpu enhances the release of new viral particles from infected cells (203, 204).

2.1.3 The Regulation of HIV-1 Gene Expression

The gene expression of the HIV-1 proteins is regulated by the long terminal repeat (LTR) promoter, which is composed of three regions: U3 (unique 3'-end), R (repeated), and U5 (unique 5'-end) (Fig. 2). These regions contain four functional elements related to the regulation of HIV-1 transcription: the transactivation response (TAR) element (nt +1 to +60) to which Tat binds, the basal core promoter (nt -80 to -1), an enhancer (nt -105 to -81), and a modulatory region (nt -454 to -106). The latter three are found within U3 region, while the TAR element is located in the R region. The modulatory region contains binding sites for numerous host cell proteins (177). However, not all of them induce HIV-1 gene expression. Apparently, the region between nt -340 and -184 contains a negative regulatory element (NRE), as deletion of this region increased HIV-1 LTR-directed transcription and viral replication (63, 193).

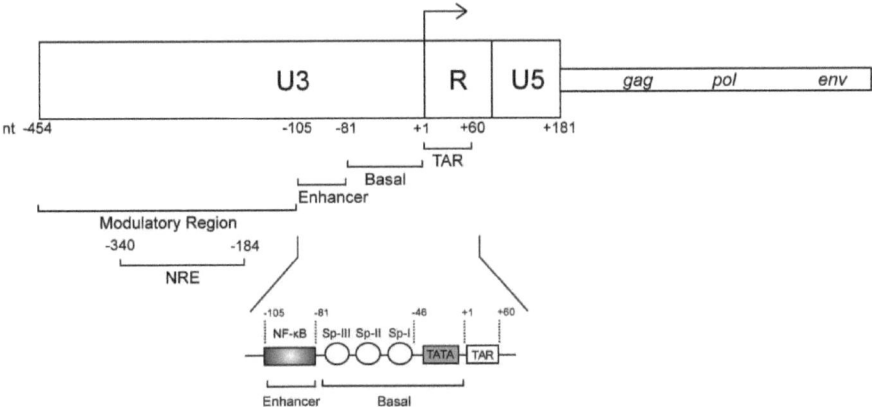

Fig. 2. Structure of the HIV-1 LTR. The U3 region contains the basal core promoter (Sp1-binding sites and TATA box), the enhancer (NF-κB-binding sites) and the modulatory region including the negative regulatory element (NRE). The TAR element is located in the R repeat. The regions encoding the viral proteins gag, pol and env are indicated. Nucleotide (nt) numbering is relative to the transcription start site nucleotide +1. Figure modified from (177).

The core promoter is composed of three binding sites for the specificity protein 1 (Sp1) and the core enhancer contains two motifs for binding of the nuclear factor-kappa B (NF-κB) (Fig. 2). Both are key elements involved in the regulation of HIV-1 transcription (63). While NF-κB is a strong enhancer of HIV-1 transcription (19), Sp1 is essential for basal transcription and Tat-mediated activation of HIV-1 (84, 220). This is concordant with the fact that Sp1 expression and activity are upregulated in activated T cells (131), which compose the primary reservoir for HIV-1 replication (217). Importantly, deletion of all three Sp1-binding sites reduces viral replication in human T cell cultures (173).

2.1.4 The HIV-1 Replication Cycle

HIV-1 infection follows a coordinated sequence of events. After the membrane fusion *via* the binding of gp120/gp41 to the CD4 receptor, the virus releases the capsid and matrix proteins into the cytoplasm of the target cell. The reverse transcription of the viral ssRNA to the proviral dsDNA occurs in the cytoplasm with the virus-associated enzyme reverse transcriptase and the ribonucloeprotein complex. In all retroviruses, reverse transcription is initiated by a tRNA whose 3' end is complementary to the so called viral primer binding site in the 5'-region of the viral RNA genome (82, 83, 186). Subsequently, the proviral DNA forms the pre-initiation complex with Vpr and the matrix protein and is transported to the nucleus (182), where the integration into the host cell genome occurs. Upon integration, viral genes are transcribed and translated to viral proteins. The assembly of viral particles occurs at the inner surface of the cytoplasmic membrane (64), followed by the budding of viral particles from the infected host cell. The cytoplasmic membrane of the infected cell forms the envelope of the virus. The viral particles maturate upon release in that the viral protease cleaves the Pol precursor protein into the reverse transcriptase, integrase and protease (37).

2.1.5 The Clinical Course of HIV-1 Infection

HIV-1 is transmitted by the transfer of contaminated blood and blood products, semen, pre-ejaculate, vaginal fluid or breast milk. The virus reaches either direct or *via* mucosa injury the blood, where it primarily infects cells of the immune system, such as $CD4^+$ T cells, macrophages (235), and dendritic cells (175). Macrophages are the primary target of HIV-1, as they are present in the mucosa; they transport the virus to the lymph nodes (223). Similarly, HIV-1 makes use of the migration of dendritic cells: it is captured by dendritic cells and delivered to the lymph nodes (190). There, HIV-1 is transmitted to $CD4^+$ T cells. The lymph node then becomes the principal site of virus production.

HIV-1 infection is associated with a progressive decrease of the $CD4^+$ T cell count and an increase in plasma viral load (Fig. 3). Three main mechanisms are involved in the decrease of $CD4^+$ T cell count: firstly, direct viral killing of infected cells; secondly, increased rates of apoptosis in infected cells; and thirdly, killing of infected $CD4^+$ T cells by $CD8^+$ cytotoxic lymphocytes that recognize infected cells (89). When $CD4^+$ T cell numbers decline below a critical level, cell-mediated immunity is lost, and the body becomes progressively more susceptible to opportunistic infections.

The stage of infection can be determined by measuring the patient's $CD4^+$ T cell count and the amount of viral particles in the blood. The course of HIV-1 infection is characterized by basically three phases (Fig. 3): acute infection, latency stage and AIDS (226). The initial incubation period upon primary infection is asymptomatic and usually lasts between two and four weeks. The acute infection (phase 1) lasts an average of 28 days and can include symptoms such as fever, lymphadenophathy, and flu-like symptoms. The latency stage (phase 2) shows few or no symptoms and can last from two weeks to twenty years and beyond. AIDS is the third and final stage of HIV infection. During this phase, HIV-1 infected patients show several symptoms of various opportunistic infections, usually leading to the death of the patient.

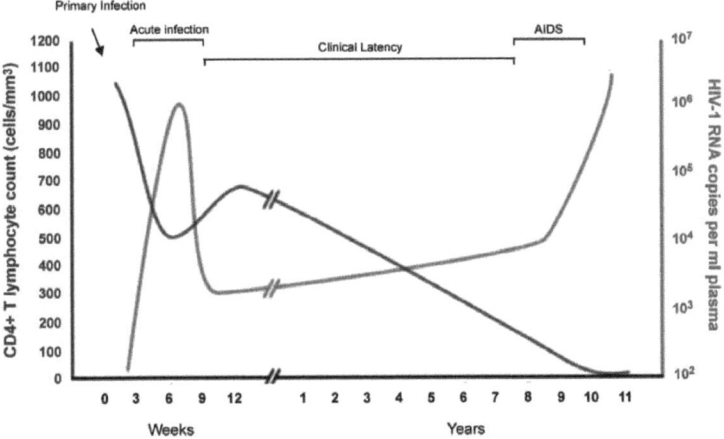

Fig. 3. Relationship between HIV-1 RNA copies (viral load) and CD4 counts over the average course of untreated HIV infection. $CD4^+$ T Lymphocyte count (cells/mm³) is indicated in dark grey, HIV-1 RNA copies per ml of plasma is indicated in light grey. The three phases during HIV-1 infection are indicated at the top. Figure modified from (170).

2.1.6 The Highly Active Antiretroviral Therapy

HIV-1 lacks proofreading enzymes to correct errors made when the viral ssRNA is reversely transcribed to the proviral dsDNA. Furthermore, the generation of new viral transcripts is mediated by the cellular RNA polymerase II, which also lacks proofreading activity. The high error rate causes the virus to mutate very rapidly, resulting in a high genetic variability of HIV-1. Some of the mutations provide a selective advantage and enable the virus to escape immune defense mechanisms (120). Furthermore, the recombinogenic properties of the reverse transcriptase enzyme and the fast turnover of virions in HIV infected individuals favour the high genetic variability of HIV-1 (44).

Combinations of several antiretroviral drugs create multiple obstacles to HIV-1 replication to keep the number of offspring low and reduce the possibility of a superior mutation. The highly active antiretroviral therapy (HAART) is a combinatorial medication of typically three different drugs for the treatment of HIV-1 infected patients (250). The antiretroviral drugs are classified in: nucleoside and nucleotide reverse transcriptase inhibitors, non-nucleoside reverse transcriptase inhibitors, protease inhibitors, integrase inhibitors, entry inhibitors, and maturation inhibitors. If a mutation arises that conveys resistance to one of the drugs being taken, the other drugs continue to suppress reproduction of that mutation.

The introduction of HAART interrupts the life cycle of HIV-1 and efficiently suppresses plasma viral loads to below the limit of detection (50 copies of viral RNA per milliliter) (164). Following the initiation of HAART, virus levels in plasma drop by 100-fold in the first 2 weeks (Fig. 4). Because HAART largely stops the new infection of susceptible cells, the decay of plasma viral loads mirrors the decay rates of productively infected cells (206). Most of the virus in the plasma of infected patients is produced by infected activated $CD4^+$ T cells, which live only a short time after infection and have a half-life ($t_{1/2}$) of one 1 day (95, 233). The entire population of circulating $CD4^+$ lymphocytes turns over every 15 days. Therefore, elimination of this cell pool is fast and results in a rapid initial decay (Fig. 4). After the clearance of the infected $CD4^+$ T cells, a second phase of decay is visible. In this phase, the level of virus in plasma decreases to the limit of detection of current assays (50 copies of viral RNA per ml of plasma) (Fig. 4). The second phase of decay results from the eradication of infected macrophages, which have a $t_{1/2}$ of about 2 weeks (181). Mathematical predictions postulated that eradication of this reservoir could be achieved within 2 to 3 years of treatment, if there were no additional populations of infected cells (178). The third phase reflects a low and chronic production of virus from a stable reservoir, which is presumably constituted by long-lived infected cells, such as dendritic cells (215) (Fig. 4). The fourth phase of decay after the onset of HAART arises from the elimination of an extremely stable and long-lived cell population, composed of latently infected $CD4^+$ T lymphocytes (30, 57, 245). The decay rate of this latent reservoir showed a striking stability with a $t_{1/2}$ of 44 months (Fig. 4). At this rate of decay, eradication of a reservoir consisting of only 10^6 cells would take 73 years (16).

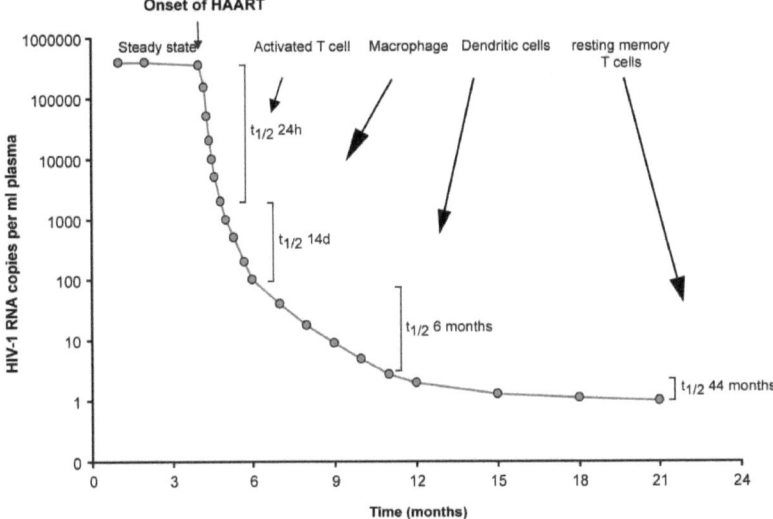

Fig. 4. Eradication of the viral reservoirs after the onset of HAART. The levels of viral RNA in plasma of infected patients before and after the initiation of antiretroviral therapy reflect the cumulative production of virus from the various cellular reservoirs and the turnover of virus-producing cells in those reservoirs. After the onset of HAART, viral replication is inhibited and plasma viral loads decay in four distinct phases. The first phase reflects virus produced predominantly from activated $CD4^+$ T cells. In the second phase, macrophages may be the main source of virus. The third phase reflects a low and chronic production of virus from a stable reservoir, constituted of infected dendritic cells. The fourth phase of decay is based on latently infected $CD4^+$ T cells. Figure modified from (215).

2.1.7 Residual Viremia and HIV-1 Latency

Although introduction of HAART stops the new infection of susceptible cells, HIV-1 replicates continuously throughout the course of infection in untreated individuals and leads to a strong decrease in the number of $CD4^+$ T lymphocytes (59). Low levels of free virus (called residual viremia) are present in the plasma even in patients responding well to the current antiretroviral regimen. Thus, HIV-1 can establish a state of latency, as cessation of HAART results in viral reappearance and high viral loads within 2 weeks (28, 40).

The molecular mechanisms of HIV-1 latency and viral reappearance have not been completely elucidated. Two possible sources explain the viral reappearance after cessation of HAART. One is that low level viral replication, below the threshold of detection, is maintained in the presence of HAART and increases without treatment (208), suggesting that HAART is not sufficient to inhibit viral replication to 100% (Fig. 5A). This theory is supported by the observation that low level viremia in patients treated with HAART can be further reduced by intensification of antiviral therapy (88). An alternative source for viral reappearance comes from the reactivation of latent

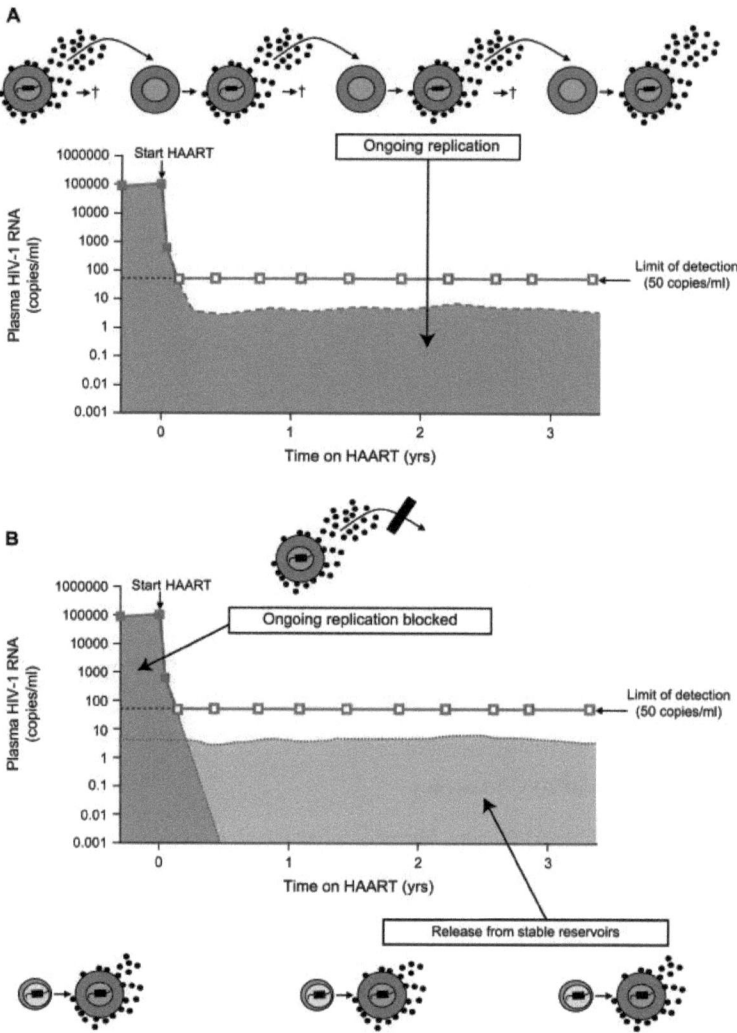

Fig. 5. Two theories for residual viremia in patients on HAART. (A) Residual viremia represents ongoing cycles of replication that continue at a lower level because of the suppressive effects of the drugs. (B) HAART stops all ongoing cycles of replication, and the residual viremia reflects release of virus from stable reservoirs such as the latent reservoir in resting $CD4^+$ T cells. Figure adopted from (206).

viruses: the integrated viral genome is transcriptionally inactive, which results in a pool of latently infected cells (Fig. 5B). *In vivo* evidence for the existence of a latent reservoir of HIV-1 infected cells was given by Siliciano and coworkers. They identified in infected patients a pool of resting memory $CD4^+$ T cells, harboring integrated provirus (27, 29). These cells arise when infected $CD4^+$

lymphoblasts (before the onset of HAART), carrying an integrated copy of the HIV-1 genome, revert back to a resting memory state. In this resting state, memory $CD4^+$ T cells are minimally permissive for virus gene expression and can survive for many years (206). Infected memory $CD4^+$ T cells begin to produce virus again after re-exposure to the relevant antigen or other activating stimuli.

Most of the studies evidence that the free virus in plasma of patients under HAART (residual viremia) is archival in character, lacks new HAART-selected mutations, and reflects earlier treatment conditions (4, 92, 121, 180). Thus, the residual viremia is nonevolving, and sensitive to drugs in the current regimen, supporting the hypothesis that the virus is produced from stable reservoirs. Whether some extremely low level of ongoing replication persists in patients on HAART remains to be confirmed more accurately by methods, which distinguish between continuing cycles of replication and continuous production of virus from stable reservoirs (206).

The existence of a latent reservoir implies that HIV-1 latency is established on a cellular level. However, the establishment of a latent reservoir seems to be an accidental consequence of HIV-1 evolution (206). Multiple mechanisms contribute to the maintenance of HIV-1 latency in resting $CD4^+$ T cells: (i) the site and orientation of provirus integration into the host cell DNA are important. Gene expression from an integrated virus is partially determined by its location in the genome and by the potential for readthrough transcription of the host gene (81, 144). Typically, this occurs when upstream transcription fails to terminate, so that the polymerase reads through into the downstream gene, thereby interfering with initiation at the downstream promoter. (ii) The local chromatin structure influences viral transcription. This has been proven by several studies showing that viral gene expression can be activated in HIV-1-infected patients after treatment with pharmacologic agents which alter the chromatin structure, such as deacetylase inhibitors (trichostatin A, sodium butyrate and pyrrole-imidazole polyamides) (184, 251). (iii) Interestingly, very highly expressed regions of the host genome are not always favored for viral integration or viral gene expression, suggesting that there might be further mechanisms governing HIV-1 gene expression, such as chromatin remodelling at the site of integration (144). (iv) Reports of increased expression and activity of DNA methyltransferase after HIV-1 infection postulate an innate genome defense against invading molecular parasites (162). (v) Further factors implicated in proviral latency include the availability or the lack of cellular transcription factors. To date, over 100 transcription factors have been shown to bind to the HIV-1 LTR promoter (177). The best studied are Sp1 and NF-κB. Both are upregulated in activated $CD4^+$ T lymphocytes (114, 131) and induce/enhance gene transcription from the HIV-1 LTR promoter (214). However, both have also been shown to recruit corepressors complexes containing histone deacetylases to the LTR promoter (105, 242), thereby maintaining HIV-1 latency. Thus, the tight regulation of cellular transcription factors might strongly influence HIV-1 gene expression. (vi) Moreover, it has recently been shown that cellular mircoRNAs contribute to HIV-1 latency by binding to the 3' untranslated region of the viral RNA, thereby leading to its degradation and inhibiting translational initiation (96). In summary, numerous changes are induced when activated T cells revert back to a resting state as memory cells. The intracellular microenvironment of resting T cells is suboptimal for HIV-1 gene expression, and profoundly silences the viral genome (134).

2.2 The Transcription Factor Sp1

2.2.1 Characteristics of Sp1

The transcription factor Sp1 (specificity protein 1) belongs to the Sp1-like/Krüppel-like family of transcription factors. It was one of the first eukaryotic transcription factors to be identified (48) and was the first transcription factor whose gene sequence was cloned (108). In pioneering studies by Tjian and coworkers, Sp1 was initially characterized as a host factor from HeLa cells that bound in a specific manner to the Simian virus 40 (SV40) early promoter at GC-rich sites and was essential for *in vitro* transcription of this promoter (49). Sp1 is a sequence specific DNA-binding protein which transactivates TATA-box-containing and TATA-less promoters. Purified Sp1 protein is 778 amino acids long (Fig. 6A) and exhibits apparent molecular masses at 95 and 105 kDa.

Sp1 has a strong DNA-binding affinity to GC-rich sites (49) with the consensus sequence 5'-G/T-**GGGCGG**-G/A-G/A-C/T-3' or 5'-G/T-G/A-**GGCG**-G/T-G/A-G/A-C/T-3' (17), but it can also bind to CT-boxes and GT-boxes, although with significant lower affinity (143). The DNA-binding activity of Sp1 has been attributed to a C-terminal region of the protein containing three Cys_2His_2 zinc fingers motifs (108, 109) (Fig. 6A) with the consensus sequence $C-X_{2-5}-C-X_3-(F/Y)-X_5-\psi-X_2-H-X_{3-5}-H$ (in the single letter amino acid code), where X represents any amino acid and ψ a hydrophobic residue (244). The first two zinc finger motifs are 23 amino acids and the third zinc finger 21 amino acids long (Fig. 6B). These sequences fold in the presence of zinc to form a compact $\beta\beta\alpha$ domain. Each finger binds a single zinc ion that is embedded between the two-stranded antiparallel β-sheet and the α-helix (Fig. 6C). The zinc is tetrahedrally coordinated between two cysteines at one end of the β-sheet and two histidines in the C-terminal portion of the α-helix (243). Although the crystal structure of Sp1 bound to its target DNA sequence is not available, data from known zinc finger structures suggest that a string of fingers wraps around the DNA.

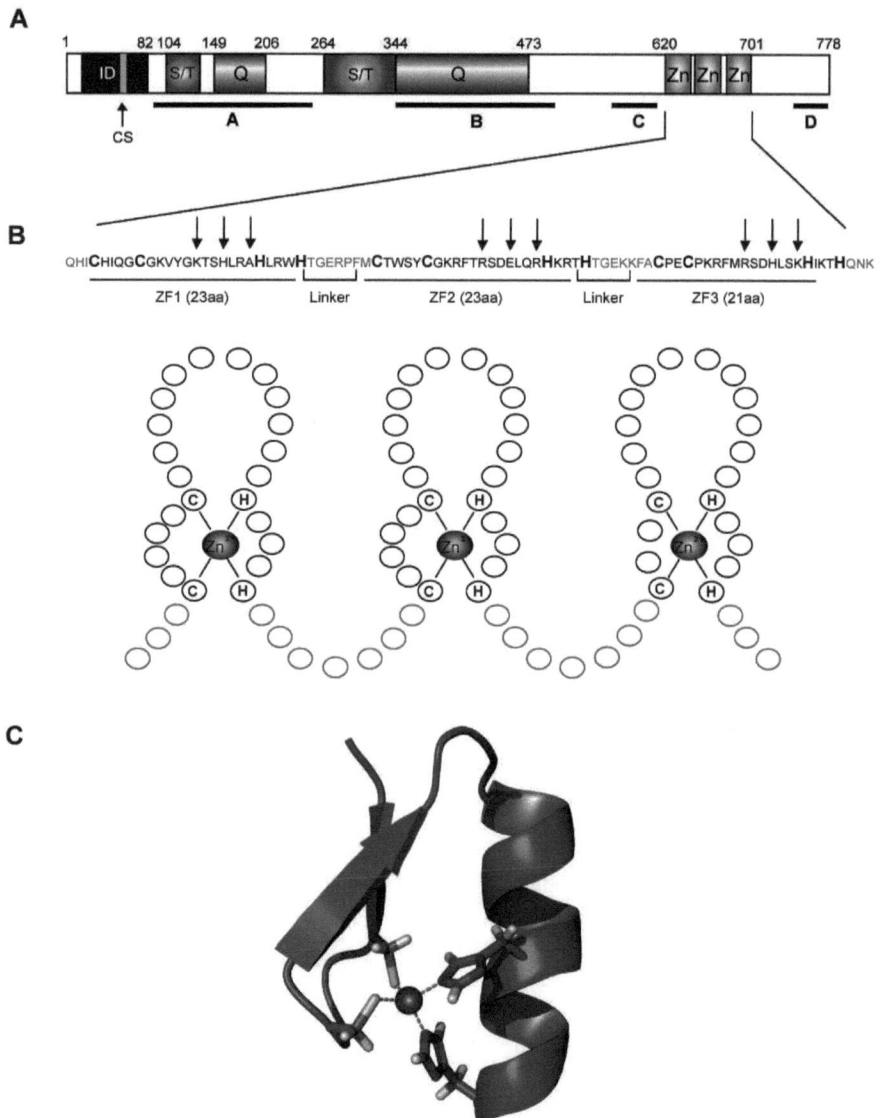

Fig. 6. Domain structure and zinc finger motifs of Sp1. (**A**) Sp1 is 778 amino acids (aa) long. Numbers above the Sp1 structure indicate the aa position in the protein. ID, inhibitory domain, S/T, serine/threonine-rich regions; Q, glutamine-rich regions; Zn, zinc finger DNA-binding domain. The transactivation domains A, B, C and D are indicated by black bars. The endoproteolytic cleavage site (CS) is indicated in light grey. (**B**) The sequence and structure of the three zinc fingers (ZF1, ZF2, ZF3) is given. The amino acid residues predicted to make the contact with DNA are indicated by arrows. The cysteins and histidines supposed to bind the zinc (green) are marked bold; the linker sequence between two zinc fingers is indicated in light grey. (**C**) Structure of a zinc finger motif. The zinc ion is coordinated by two histidine and two cysteine amino acid residues. Figure (zinc finger) made by Thomas Splettstoesser.

2.2.2 Transcriptional Activation by Sp1

Sp1 contains four domains (A, B, C and D) involved in transcriptional activation (Fig. 6A): the domains A and B contain glutamine rich motifs and are the two strong transactivation domains (36). The domain C is highly charged and possesses only weak transactivation potential, while domain D completely lacks own transactivation potential, but is necessary for synergistic activation (241). At the extreme N-terminus, Sp1 contains an inhibitory domain, which can be cleaved off the molecule due to an endoproteolytic cleavage site at position 54 (219).

Sp1 can directly interact *via* hydrophobic residues in the transactivation domains A and B with components of the basal transcription machinery, which is essential for Sp1-driven transcription. Direct interactions of Sp1 with the TATA-binding protein (TBP) and TBP-associated factors (TAFs) have been shown (53, 66, 94). Furthermore, Sp1 efficiently recruits TBP/TFIID to promoters and stimulates transcription initiation *via* correct positioning of the initiation complex, which is mainly important for the activation of TATA-less genes (125).

Sp1 activates gene expression *via* several mechanisms and numerous reports proved that different domains of Sp1 are required for distinct functions (35, 36, 127, 174).

a) Activation through one or more independent sites

Sp1 was discovered due to its ability to activate the SV40 early promoter, which contains six GC-boxes (49). Although the binding affinity of Sp1 for these different sites varied considerably, no evidence for cooperative interactions was found (65) (Fig. 7A). The transactivation potential for activation by binding to one or several independent sites resides in the domains A and B, while domains C and D are dispensable (35). Sp1 functions as a monomer or multimer at a single site: it binds to a single GC box as a monomer and then builds higher-order complexes (tetramers) by direct protein-protein interactions that do not appear to involve additional contacts with the DNA (174).

b) Synergistic activation by multimerization of Sp1

Sp1 can multimerize to synergistically activate target promoters containing at least two binding sites for Sp1 (located adjacent or distal to each other) (Fig. 7B). The synergistic activation by Sp1 multimers is dependent on the specific promoter, as not each promoter containing GC-boxes displays synergism, and on the transactivation domains A, B and D (174). The formation of these complexes may generate an activation surface that can interact more efficiently with the general transcription machinery and thereby activate transcription to a higher level.

c) Superactivation by Sp1

Tjian and coworkers observed that a DNA-binding deficient mutant of Sp1 that retains glutamine-rich domains was still able to interact with DNA-bound wild-type Sp1 and enhance its ability to activate transcription (Fig. 7C). Thus, one Sp1 molecule is able to interact with a DNA-bound Sp1 molecule, thereby mediating superactivation of the targeted promoter (35).

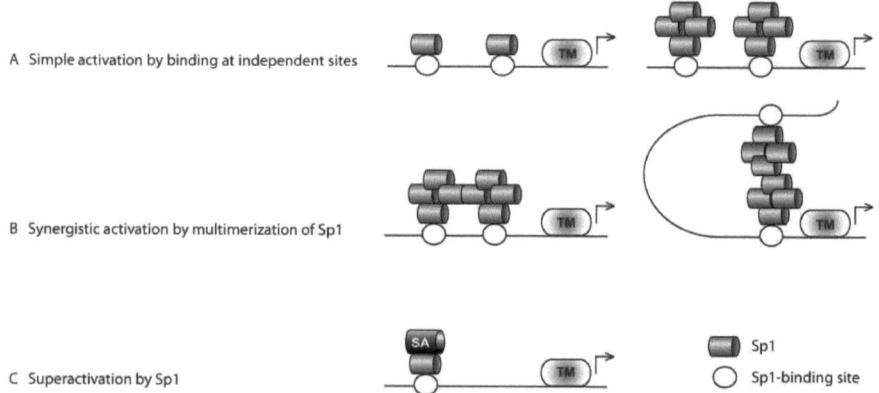

Fig. 7. Model for the different activation mechanisms by Sp1. (A) Sp1 monomers or multimers can bind to a single site and interact with the general transcription machinery (TM) to activate a low level of transcription. **(B)** Multimers (tetramers) of Sp1 bound at adjacent or distal sites interact to form higher-order complexes and thereby activate transcription synergistically. **(C)** One Sp1 molecule bound to the DNA can interact with another Sp1 molecule (SA), which does not bind to DNA but superactivates Sp1-mediated transcription. Figure modified from (174)

It is presently not solved, how the ability of Sp1 to activate gene expression either by acting in a higher order complex or by binding as a monomer to its target sites might be modulated. Tjian and colleagues have shown that placing Sp1 sites upstream of Sp1-triggered promoters does not necessarily lead to synergistic activation by Sp1 (174). Thus, synergistic activation and superactivation by Sp1 may depend on the promoter context.

2.2.3 Regulation of the Transcriptional Activity of Sp1

The ubiquitous transcription factor Sp1 is not only a constitutive activator of housekeeping genes and other TATA-less genes, but also an important regulator of inducible gene expression involved in cell growth, cell proliferation, cell cycle progression, angiogenesis and tumorigenesis. Therefore, the transcriptional activity of Sp1 is tightly regulated: first, by its expression level, second, by posttranslational modifications and third, by the interaction with other proteins.

2.2.3.1 Regulation of Sp1 Expression Level

Although Sp1 is ubiquitous, its expression is subject to specific regulation. Expression of Sp1 differs between the cell types during development: Sp1 is downregulated in several fully differentiated cells, but highly expressed in developing cells during differentiation (194). The level of Sp1 is not only regulated on RNA level, but also on protein level. Sp1 is a substrate for proteolytic cleavage and undergoes proteasome-dependent degradation under conditions of nutrient starvation (80). Cleavage of Sp1 can also occur by caspase-3-like protease (189), cathepsin-like protease (167), and myeloblastin protease (185). Cleavage of Sp1 by myeloblastin protease results in a truncated form of about 30 kDa, which lacks the transactivation domains, but contains the C-terminal zinc fingers and is able to bind to DNA (185). This truncated form might act as a transcriptional inhibitor, as it competes with full-length Sp1 for the same binding sites.

2.2.3.2 Regulation by Posttranslational Modifications

The multitude of different regulatory functions of Sp1 is explained in part due to posttranslational modifications, which include phosphorylation (25), O-GlcNAcylation (101, 191, 247), ubiquitinylation (1, 212), sumoylation (211, 234), poly(ADP-ribosyl)ation (254, 255), and acetylation (97, 222).

a) Phosphorylation of Sp1

Sp1 is phosphorylated by at least nine kinases. Five phosphorylation sites on Sp1 have been confirmed: Ser52 (quoted as Ser59 in the original article), Ser213 (quoted as Ser131), Thr446 (quoted as Thr453), Thr661 (quoted as Thr579) and Thr732 (quoted as Thr739). A further phosphorylation site at Thr348 (quoted as Thr266 in the original article) has been suggested, although direct evidence was not presented (259). Residues Ser213, Thr348 and Thr446 reside in the two transactivation domains A and B, and residue Thr732 in the domain D (Fig. 8). Inactivation of these sites decreased the ability of Sp1 to activate certain promoters (25). It has therefore been suggested that phosphorylation plays an important role in modulating the interaction of Sp1 with partner proteins at the promoter (25). In contrast to this, phosphorylation of the residue Thr661 located in the second zinc finger in the DNA-binding domain of Sp1 (Fig. 8) resulted in a decrease in the DNA-binding activity of Sp1 (3). Thus, phosphorylation of Sp1 can have positive and negative effects on Sp1 functions such as DNA-binding and promoter activation (26, 58, 145). As the signals, pathways, kinases and phosphatases that regulate Sp1 phosphorylation are extremely diverse, Sp1 can be phosphorylated at various sites or combinations of sites, leading to a wide range of changes in Sp1 function.

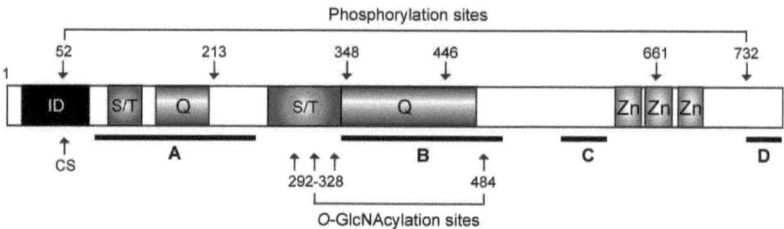

Fig. 8. Schematic representation of phosphorylation and *O*-GlcNAcylation sites in Sp1. ID, inhibitory domain, S/T, serine/threonine-rich regions; Q, glutamine-rich regions; Zn, zinc finger DNA-binding domain. The transactivation domains A, B, C and D are indicated by black bars. The endoproteolytic cleavage site (CS) is indicated. Arrows indicate amino acids which are targets of phosphorylation and *O*-GlcNAcylation. Residues for both modifications are located in transcriptionally important regions: inhibitory domain, transactivation domains A, B and D, and zinc finger motif.

b) O-GlcNAcylation of Sp1

Two decades ago, Tjian and coworkers showed that Sp1 bears at least eight *O*-GlcNAc sites, and that the *O*-GlcNAcylated form of Sp1 was more active than the non-*O*-GlcNAcylated protein (101). Several studies have confirmed that *O*-GlcNAcylation enhances Sp1-dependent transcription in some experimental conditions (68, 151, 152). In addition, *O*-GlcNAcylation protects Sp1 from proteolysis and controls the stability of Sp1 (80, 219). In contrast to this, *O*-GlcNAcylation of Sp1 has also been described to inhibit its activity. This has been attributed to the inhibition of hydrophobic interactions with other transcription factors: *O*-GlcNAcylation of Ser484 (Fig. 8) in the transactivation domain B inhibits the dimerization of Sp1 and its interaction with the TFIID-associated transcription factors like TAF$_{II}$110 (*Drosophila* homolog of human TAF4) (191, 247). Several additional *O*-GlcNAc sites are located in the serine/threonine-rich region of mouse Sp1 between the amino acids 302-337, corresponding to amino acids 292-328 in the human Sp1 (Fig. 8). *O*-GlcNAc modifications within this region interrupt a known interaction between Sp1 and Oct1, and inhibit the cooperative activation of target genes (122).

Evidence has been obtained that the differential effect of *O*-GlcNAcylation on Sp1 activity may strongly depend on the targeted promoter and the respective transcription complexes involved in the regulation of the specific promoter. Furthermore, since *O*-GlcNAcylated proteins are phosphoproteins and *O*-GlcNAcylation occurs at serine and threonine residues which are also targets of phosphorylation, an inverse relationship between Sp1 phosphorylation and *O*-GlcNAcylation has been proposed (209).

2.2.3.3 Regulation by Interaction Partners

Sp1 regulates highly specific the expression of different target genes *via* interaction with other transcription factors and with chromatin remodeling complexes. The high variety of interaction partners enables Sp1 to regulate the expression of numerous genes, among them housekeeping genes and other TATA-less genes, but also inducible genes involved in cell growth, cell proliferation, cell cycle progression, angiogenesis and tumorigenesis (241).

Sp1 transactivates synergistically together with a variety of transcription factors, such as p65/RelA, Oct-1 and AP-2, oncogenes (retinoblastoma protein [RB], Smad4), and tumor suppressors (p53, cyclin D1, v-Jun, c-Myc) (241). Moreover, a lot of transcription factors (such as c-Jun, AP-2, E2F-1, Smad2, Smad3 and Smad4) superactivate the Sp1-mediated transactivation by binding to DNA-bound Sp1, but not to DNA. However, interaction of Sp1 with other transcription factors does not always increase Sp1-regulated transactivation. The direct protein-protein interaction between Sp1 and the oncogene VHL (Van-Hippel-Lindau) blocks the DNA binding activity of Sp1. The same effect can be observed when Sp1 interacts with MDM2 (murine double minute 2).

Sp1 is further known to interact with different proteins involved in chromatin remodeling. On one side, Sp1 is able to recruit proteins responsible for the decondensation of chromatin, such as the chromatin remodeling complex SWI/SNF and the histone acetyltransferase (HAT) CBP/p300 (10). On the other side, Sp1 can recruit corepressors to the promoter, like DNA methyltransferase-1, histone deacetylases (HDAC)-1 and 2, thereby blocking the spreading of heterochromatin. Additionally, the extreme N-terminus of Sp1 harboring the inhibitory domain (ID, Fig. 8) suppresses transactivation by recruiting corepressors such as SMRT (silencing mediator of retinoid and thyroid receptor), NCoR (nuclear hormone receptor corepressor) and BCoR (BCL-6 interacting corepressors) to the promoter (138, 211).

In general, Sp1 appears to be a very versatile interaction partner for other proteins, of which the majority is tightly regulated in different tissues and during the cell cycle. The transactivation ability of Sp1 dependents on the cellular conditions, the promoter context and the availability of interacting proteins. This explains the highly specific regulation of a large number of different promoters.

2.3 Posttranslational Modification of Proteins with *O*-GlcNAc

Initially described in 1984 as a terminal modification of lymphocyte cell-surface proteins, it has been shown that *O*-GlcNAc can also exist as a single monosaccharide *O*-glycosidically linked to proteins inside the cell (227). The *O*-glycosidic linkage of GlcNAc to proteins is a highly dynamic and reversible posttranslational modification at serine and threonine residues of target proteins. This suggested that *O*-GlcNAc modification may have regulatory function. *O*-GlcNAcylation differs from other glycosylation events in that it occurs in the cytosol and the nucleus rather than in the Golgi apparatus or the endoplasmic reticulum (227).

Over the intervening years, myriad cytoplasmic and nuclear proteins have been identified to be modified at serine and threonine residues by *O*-GlcNAc (*O*-GlcNAcylation), including transcription factors, cytoskeletal components, metabolic enzymes, signaling components, chromatin proteins, kinases, oncoproteins and others (32). As yet, there have been no reports of *O*-GlcNAc in prokaryotes, suggesting that *O*-GlcNAcylation is restricted to eukaryotes.

2.3.1 The Hexosamine Biosynthesis Pathway

The substrate for protein *O*-GlcNAcylation is UDP-GlcNAc. UDP-GlcNAc is synthesized from glucose *via* the hexosamine biosynthetic pathway (HBP). About two to three percent of the incoming glucose enters the HBP. The first and rate-limiting step in this pathway, the conversion of fructose-6-phosphate to glucosamine-6-phosphate and the concomitant conversion of glutamine to glutamate, is catalyzed by the enzyme glutamine:fructose-6-phosphate amidotransferase (GFAT, Fig. 9). Glucosamine-6-phosphate is further converted trough several steps to UDP-GlcNAc, which can be used for *O*-GlcNAc modification of proteins or for the generation of other UDP-sugars (238). Thus, glucose flux through the HBP increases *O*-GlcNAcylation of proteins. Therefore, *O*-GlcNAcylation is regulated by the cellular levels of the enzymes cycling *O*-GlcNAc, OGT and *O*-GlcNAcase, and by the availability of UDP-GlcNAc (253). Interestingly, while *O*-GlcNAc levels are very responsive to the change in UDP-GlcNAc levels, complex *O*- and N-linked glycosylation levels are not.

Compounds like glucosamine (GlcN), streptozotocin (STZ), O-(2-acetamido-2-deoxy-D-glucopyranosylidene)amino-N-phenylcarbamate (PUGNAc), and 2-deoxyglucose (2-DG) enhance *O*-GlcNAcylation of proteins either by increasing the availability of UDP-GlcNAc or by inhibiting the enzyme *O*-GlcNAcase (79, 113, 192).

Interestingly, elevated O-GlcNAc levels have been linked to insulin resistance and diabetes (18, 159, 176). One molecular mechanism for how O-GlcNAc exerts this effect relies on the O-GlcNAcylation of signaling molecules such as insulin receptor, insulin receptor substrate (IRS) and Akt, thereby inhibiting their insulin-dependent activation (5, 176) (Fig. 9). A second molecular mechanism for O-GlcNAc-mediated insulin resistance has been attributed to the O-GlcNAcylation of the insulin-responsive glucose transporter Glut4 and its impaired translocation to the plasma membrane (7, 172). Moreover, OGT is also regulated in the insulin signaling pathway. Recently, it has been shown that OGT is activated by insulin signaling: insulin stimulates the catalytic activity of OGT and a partial translocation of OGT from the nucleus to the cytoplasm, resulting in increased O-GlcNAc modification of OGT and its substrates (239).

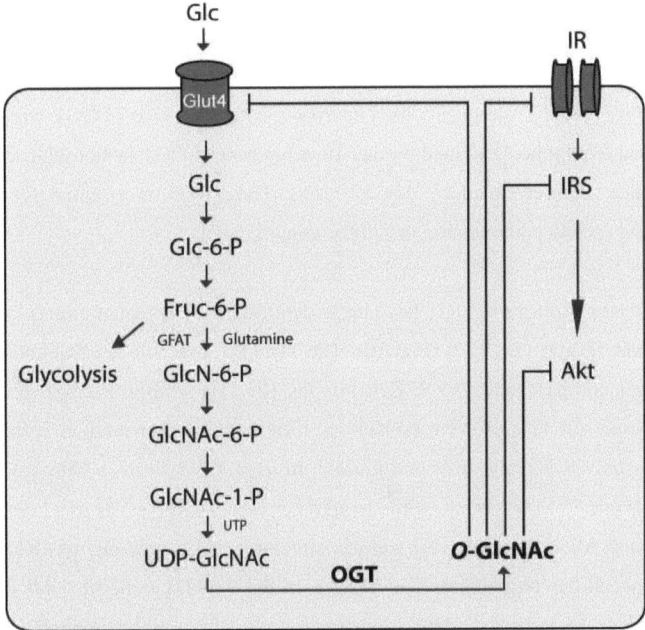

Fig. 9. The hexosamine biosynthesis pathway. The incoming glucose (Glc) enters the HBP, where it is metabolized *via* several steps to UPD-GlcNAc. Increased flux through the HBP results in greater UDP-GlcNAc levels, upon which O-GlcNAc transferase (OGT) is highly dependent. Increased O-GlcNAcylation of Glut4 impairs its translocation to the membrane and establishes a feedback for glucose homeostasis. O-GlcNAcylation of the insulin receptor (IR) and its substrate (IRS) as well as Akt has been linked to insulin resistance due to impaired insulin-dependent signaling. Figure modified from (210).

2.3.2 The Cycling Enzymes Involved in *O*-GlcNAc Modification of Proteins

O-GlcNAcylation is catalyzed by *O*-GlcNAc transferase and reversed by *O*-GlcNAc hexosaminidase (100).

2.3.2.1 O-GlcNAc Transferase

OGT is the sole enzyme known to mediate *O*-GlcNAc modification of proteins. All metazoans studied to date, including plants, contain a highly conserved OGT (128, 147), while neither enzyme nor *O*-GlcNAc itself is found in prokaryotes or budding yeast, suggesting that *O*-GlcNAcylation evolved with multicellular organisms.

In humans, the *OGT* gene is mapped to the X chromosome region q13. Knock-out in mice led to embryonic lethality (205), while mice with a neuron-specific knockout of OGT were smaller, failed to develop normal locomotion and died by day 10. Knockout of OGT in fibroblast cell line induced growth senescence; the cells died by day 12 (168). Thus, OGT is essential for proper cellular function including growth and response to cellular stimuli (210).

So far, three different isoforms of OGT have been identified, with apparent molecular masses of 116 kDa, 103 kDa and 78 kDa (78, 137) (Fig. 10). The 116 kDa molecule is ubiquitously expressed in nucleocytoplasmic compartments (ncOGT), while the 103 kDa isoform is targeted to mitochondria (mOGT). The small 78 kDa isoform (sOGT) is differentially expressed in certain tissues with detectable levels only in kidney, liver and muscle. In liver, OGT forms a heterotrimer of two 116 kDa subunits and one 78 kDa subunit, while in all other tissues OGT exists as a homotrimer of three 116 kDa isoforms. All isoforms have a similar structure: the N-terminus of OGT appears to be important for subcellular targeting and is lacking in the 78 kDa molecule. All isoforms further contain a tetratricopeptide repeat (TPR) domain, followed by a linker region, and a C-terminal highly conserved catalytic domain (Fig. 10). TPRs are 34-amino acid repeats, which modulate the interactions of OGT with its protein substrates. The number of TPRs can vary: ncOGT contains twelve, mOGT nine and sOGT three TPRs (78). The crystal structure of the human TPR domain of OGT has been analyzed and shows similarities to importin-α, in that the TPR domain clearly forms an elongated, flexible scaffold of numerous electrostatic surfaces, allowing multiple protein-protein interactions (106).

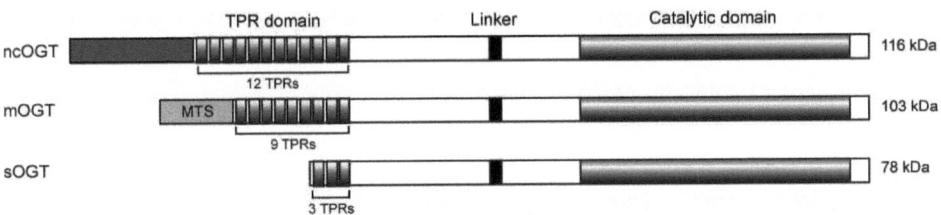

Fig. 10. Structure of OGT isoforms. OGT contains 3 to 12 TPRs, a linker domain, followed by the well conserved, C-terminal catalytic domain. ncOGT and mOGT contain distinct N-termini, indicated by dark and light grey, respectively. The N-terminus of mOGT contains a mitochondrial targeting sequence (MTS). All isoforms contain identical catalytic domains.

The activity of OGT can be regulated by several mechanisms: first, by the UDP-GlcNAc levels in the nucleus and cytoplasm. Kinetic studies revealed that OGT possesses three different binding affinities for UDP-GlcNAc, depending on the concentration of the substrate in the environment (129). As the cellular levels of UDP-GlcNAc are known to vary dramatically between tissues (37), this finding strongly suggests that OGT activity is regulated by the UDP-GlcNAc levels in the cytosol and nucleus. Second, UDP-GlcNAc levels modulate the affinity of OGT for individual glycosylation sites differentially, meaning that peptides that are poor acceptors substrates at low concentrations of UDP-GlcNAc would be better acceptors at higher UDP-GlcNAc concentrations. Third, OGT can be regulated by its multimerization state, which alters its binding affinities for UDP-GlcNAc: the lowest Km value of the homotrimer for UDP-GlcNAc is 10-fold higher than that reported for the p78/p116 heterotrimer found in liver, kidney and muscle (129). Moreover, OGT is regulated by posttranslational modifications such as *O*-GlcNAcylation and tyrosine phosphorylation (128, 239). As yet, the sites of modifications have not yet been mapped. However, at least one site of *O*-GlcNAcylation lies within the C-terminal half of the protein and the site(s) of tyrosine phosphorylation map to the TPRs 7 to 9 (129).

2.3.2.2 O-GlcNAc Hexosaminidase

O-GlcNAcase was first characterized by Hart and coworkers (43), and was found to be identical to MGEA5, a putative hyaluronidase (91). *O*-GlcNAcase is a neutral hexosaminidase with substrate specificity toward GlcNAc and the only protein known to remove the *O*-GlcNAc modification *via* its N-terminal β-N-acetylglucosaminidase domain (Fig. 11). In addition, *O*-GlcNAcase has a C-terminal histone acetyltransferase (HAT) domain and can acetylate all four core histones, similar to other HATs like p300 and CBP (225). Due to its bifunctionality, the enzyme has been termed nuclear/cytoplasmic *O*-GlcNAcase and acetyl transferase (NCOAT). The gene maps to chromosome 10q24 and this locus has been implicated in Alzheimer's disease (91). The transcript is expressed in every human tissue examined, but highest in the brain, placenta and pancreas. The enzyme has a nucleocytoplasmic distribution, with predominant localization in the cytoplasm.

Full-length NCOAT is a 916-amino acid protein having a predicted molecular weight of 103 kDa (Fig. 11), but the protein migrates at 130 kDa in gel electrophoresis (62). Three splice variants of NCOAT were identified, two of them lacking parts of the N-terminal domain containing the hexosaminidase activity (225) and one lacking the C-terminal HAT domain due to an alternative stop codon. In addition, it has been shown that NCOAT is a substrate for caspase 3 cleavage and that the processed product retains *O*-GlcNAcase activity (237). Thus, the activities of the two different domains do not rely on each other. The functional independence of both domains and the cleavage by caspase 3 have been proposed to be an additional mechanism to control NCOAT activity. Furthermore, posttranslational modification may also correlate with the activation of the enzymatic domains, as bacterially expressed NCOAT totally lacks activity, but recovers when incubated with a mammalian whole cell lysate (225).

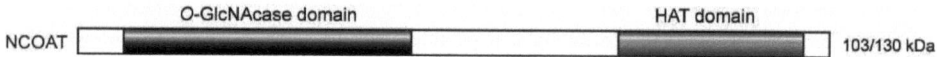

Fig. 11. Structure of NCOAT. The *O*-GlcNAcase domain is located at the N-terminus, the HAT (histone acetyltransferase) activity resides in the C-terminus of the enzyme. The predicted molecular weight is 103 kDa, but the enzyme migrates at 130 kDa.

2.3.2.3 Interplay Between NCOAT and OGT at the Chromatin

Work from Kudlow and colleagues has provided evidence that OGT is recruited to promoters by corepressors (13, 248). Corepressors are proteins which are unable to bind DNA by themselves, but decrease gene expression by binding to a transcription factor. Usually, corepressors repress transcription by recruiting HDACs to the promoters. The latter catalyze the removal of acetyl groups from the lysine residues of histones, leading to an increase of the positive charge on histones, which strengthens the interaction between histones and DNA (136). This results in heterochromatic DNA, which is less accessible to transcription. OGT has been shown to interact with mSin3A, NCoR (nuclear receptor corepressor 1), and SMRT (silencing mediator for retinoid and thyroid-hormone receptors) along with HDAC1 (13, 248). The interaction with corepressors targets OGT to sites of transcriptional repression and enables it to contribute to the corepression of eukaryotic gene expression by *O*-GlcNAcylation of the RNA polymerase II and associated factors. This is confirmed by the observation that repressed promoters are hyper-*O*-GlcNAcylated (240). Thus, *O*-GlcNAcylation is connected to the modulation of the transcriptional activity, along with alteration of the chromatin structure by HDACs.

Accordingly, activation of gene expression requires the converse enzymes, an *O*-GlcNAcase and HAT, to reverse the protein modification involved in the repression. Transcriptional activation is associated with relaxed euchromatic DNA, which is achieved by the acetylation of conserved lysine residues on histones by HATs. Especially acetylation of histone H4 appears to play a primary role in the structural changes at the promoter, mediating enhanced binding of transcription factors to their recognition sites within nucleosomes (232).

Recently it has been shown that OGT directly associates very tightly with NCOAT *in vivo* and *in vitro* (240). As NCOAT resides with OGT and corepressors in the repression complex, it is strategically placed to rapidly reverse both repressive modifications: *O*-GlcNAcylation and histone deacetylation. The tight regulation of the enzymatic activities of NCOAT and the association of opposing enzymes, NCOAT and OGT, in one complex provides the necessary properties for dynamic regulation of gene expression in response to environmental stimuli and glucose homeostasis: reversibility, rapidity, and thoroughness.

2.3.3 Diverse Regulation of Protein Function by *O*-GlcNAc

Numerous *O*-GlcNAc-modified proteins have been identified to date and are summarized in Table 1. Analysis of the *O*-GlcNAc proteome has revealed a diverse set of proteins involved in numerous cellular functions, such as carbohydrate metabolism, signaling, nuclear transport, transcription/translation, cell cycle, and stress response (Fig. 12). The broad variety of *O*-GlcNAc-modified proteins explains how a simple monosaccharide can modulate (a) transcription, (b) translation, (c) nuclear transport, and (d) protein degradation (253).

Table 1. *O*-GlcNAc-modified proteins

Nuclear pore
Nup153, 214, 358
Nup180
Nup54, 155
p62

Transcription factors
AP-1
CTF
c-Fos and c-Jun
c-Myc
CREB
ELF-1
Enhancer factor 2D
Estrogen receptor
h-Catenin
Hepatocyte nuclear factor 1
Human C1 transcription factor
NF-κB
Oct1
P107
p53
Pancreas-specific transcription factor
PDX-1
PAX-6
Plakoglobin
RB
RNA polymerase II
Serum response factor
Sp1
v-erbA
YY1

RNA-binding proteins
40S ribosomal protein s24
EF1
Eukaryotic initiation factor 4A1
Ewing-sarcoma RNA-binding protein

Phosphatases and kinases
Adapter proteins insulin receptor substrate 1, 2
CKII
GSK-3
Nuclear tyrosine phosphatase p65
PI3-kinase

Cytoskeletal proteins
Actin-based B and 4.1
Ankyrin
Chaperones
Cofilin
Dynein LC1
E-cadherin
HsC
HsP 27, 70, 90
Intermediate keratins 8,13, 18
MAP 2 and 4
Myosin
Neurofilaments H, M, L
Synapsin
Talina
Tau
Vinculin
α-Tubulin

Enzymes
CRMP-2
Enolase
eNOS
Eukaryotic peptide chain initiation factor-2
GAPDH
Glycogen synthase
OGT
Phosphoglycerate kinase
Pyruvate kinase
UCH-L1
UDP-glucose pyrophosphorylase

Other
Adenovirus type 2 and 5 fiber proteins
Annexin 1
AP-3 and -180
Gaba-receptor interacting protein-1 splice variants
Glut-1
h-APP
h-Synuclein
Nucleophosmin
p67
Peptidyl prolylisomerase
Phosphatase-2a inhibitor
Piccolo
Proteasome component C2
Q04323, UCH homolog
Rho GDP-dissociation inhibitor 1

Viral proteins
Baculovirus GP41
HCMV UL32 (BPP) tegument protien
NS26 rotavirus protein
SV-40 large T-antigen
Adenovirus type 2 and 5 fiber protein
Viral basic phosphoprotein

Additional proteins identified by GlcNAc incorporation (213)
14-3-3
Ade5
Aminopeptidase
Calreticulin
Capulet
Catalase
CG10417 Protein S/T phophatase
CG1129 Mannose-1-phosphate guanylyltransferase
CG2767 Alcohol dehydrogenase
CG32626 Purine base metabolism
CG6180 Kinase inhibitor
Chd64
EGFR-3
Glutatione-S-transferase 1
Myo-inositol-1-phosphate synthase
Prolyl oligopeptidase
Proteasome alpha 4 and 6 subunit
Protein disulfide isomerase
Ran
Regucalcin
Slow superoxide dismutase
Thioredoxin peroxidase 1
Ubiquitin activating enzyme

Nup, nuclear pore complex proteins / nucleoporins; Ap-1, activator protein 1; CTF, CCAAT-binding transcription factor; CREB, cyclic AMP response element-binding protein; ELF-1, E74-like factor 1; ER, estrogen receptor; NF-κB, nuclear factor kappa B; Oct1, octamer-binding protein; PDX-1, pancreatic/duodenal homeobox-1 protein; PAX-6, paired box protein; RB, retinoblastoma protein; Sp1, Specificity protein 1; YY1, yin yang-1; EF1, elongation factor 1; CKII, Cyclin kinase II; GSK-3; Glycogen synthase kinase 3, PI3-kinase, phosphoinositol 3-kinase; MAP, microtubule associated protein; HsP, heat shock protein; HsC 70, heat shock cognate protein 70; eNOS, endothelial nitric oxide synthase; GAPDH, Glyceraldehyde-3-phosphate dehydrogenase; OGT, O-GlcNAc Transferase; UCH-L1, Ubiquitin carbxy hydrolase L1; AP-3 and -180, clathrin assembly protein 3 and 180; h-APP, human amyloid precursor protein; Glut-1, glucose transport protein; HCMV UL32 (BPP), human cytomegalovirus unique left 32 basic phosphoprotein; Chd64, Calponin-like protein; EGFR-3, epidermal growth factor receptor type 3. List compiled from (146).

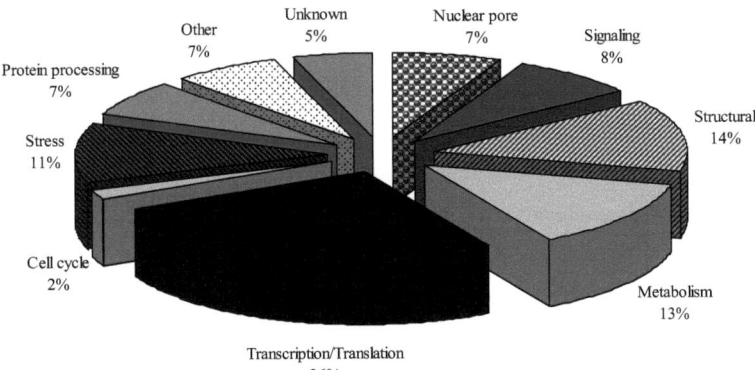

Fig. 12. Pie diagram of the *O*-GlcNAc proteome. The known *O*-GlcNAc substrates are divided into groups according to their function. Proteins involved in transcription/translation constitute the largest group, followed by the groups encompassing proteins involved in structural functions and carbohydrate metabolism. Diagram compiled from (146).

a) Control of Gene Transcription

Classification according to the annotated function of the *O*-GlcNAc-modified proteins reveals that most of them are involved in gene regulation (transcription/translation, 26%, Fig. 12). This reflects the major task of *O*-GlcNAc in these processes. *O*-GlcNAc plays a pivotal role in the regulation of gene expression *via* modification of the largest RNA polymerase II subunit on its C-terminal domain (CTD) (33, 119). Subsequent to initiation, the *O*-GlcNAc moieties on the CTD are removed, allowing phosphorylation to occur. During transcriptional elongation, the CTD is extensively phosphorylated (135), suggesting that *O*-GlcNAcylation is involved in the assembly of the pre-integration complex (85). Furthermore, almost every RNA polymerase II associated transcription factor that has been examined is *O*-GlcNAcylated (101, 102) (Table 1).

b) Control of Protein Synthesis

The eukaryotic initiation factor 2 (eIF-2) is a critical target for the regulation of protein translation. Phosphorylation of eIF-2 inhibits protein synthesis. The activity of eIF-2 is regulated by a 67 kDa protein, in that association of p67 protects eIF-2 from phosphorylation, thereby initiating translation (38). Interestingly, p67 is a glycoprotein containing several *O*-GlcNAc moieties. Removing the *O*-GlcNAc moieties from p67 results in degradation and allows kinase-dependent phosphorylation of eIF-2 (39). Thus, under conditions of nutrient deprivation, protein synthesis is downregulated.

c) Control of Nuclear Transport

A correlation between *O*-GlcNAcylation and the nuclear accumulation of transcription factors has been described (74). Direct evidence for *O*-GlcNAc playing a role in nuclear transport was given from early studies on nuclear pore proteins, which were among the first identified O-GlcNAc proteins (95). It has been described that microinjection of an antibody binding to the O-GlcNAc residues on the nuclear pore complex inhibits protein import into the nucleus (55). Recent studies have provided evidence, that O-GlcNAcylation of Sp1 modulates its nuclear translocation (151, 152), confirming the involvement of O-GlcNAc in the modulation of protein localization.

d) Impact on Protein Degradation

Several studies have reported that O-GlcNAc modification of proteins such as the transcription factor Sp1 and the eIF-2-associated protein p67 stabilizes them towards proteosomal degradation (20, 80). Removal of O-GlcNAc results in increased proteasome susceptibility. As yet, it is not known whether O-GlcNAc-prevented degradation occurs due to blocked phosphorylation in a sequence signalizing degradation (like PEST sequences) or O-GlcNAc removal alone serves as degradation signal. An additional involvement of O-GlcNAc in the regulation of protein stability was provided by Kudlow and coworkers. They observed that O-GlcNAcylation of the 26S proteasome prevents proteolysis by inhibiting the ATPase activity of the proteasome (257).

About 25% of the cellular energy is used for protein synthesis and processing (99). It is conceivable that the regulation of the transcription/translation machinery is coupled to energy availability. Thus, O-GlcNAc may serve as a nutrient sensor to couple the metabolic status to cellular processes, such as gene expression, signal transduction and protein degradation.

2.3.4 The Yin-Yang Hypothesis

O-GlcNAc-modified proteins are usually phosphoproteins, which form reversible multimeric complexes with other peptides *via* phosphorylation-dependent associations. Three features suggest that O-GlcNAc plays an antagonistic role to protein phosphorylation: (i) similar to phosphorylation, O-GlcNAc is highly dynamic with rapid cycling in response to cellular signals or stages (116); (ii) O-GlcNAc is reciprocal with phosphorylation on some well-studied proteins, such as RNA Pol II (33), c-Myc (24), SV-40 large T-antigen (161) and estrogen receptor-β (23); and (iii) O-GlcNAcylation and phosphorylation occur on same or adjacent serine or threonine residues on the protein backbone (78, 87). The dynamic interplay between O-GlcNAcylation and phosphorylation has led to the development of the "yin-yang" hypothesis, in which O-GlcNAc and phosphorylation compete for the same site or region on the peptide, thereby modulating protein functions. Data supporting this hypothesis came from Hart and coworkers. They showed that glycosylation prevents (33) or reduces (31) phosphorylation of a peptide substrate in an *in vitro* kinase assay. Further reports demonstrated that the global level of protein *O*-GlcNAcylation inversely correlates with the phosphorylation level (73, 139). These data imply a complex model of protein regulation, where diverse posttranslationally modified forms of a protein can exist (Fig. 13), each of them with different functions and/or activity (252). The regulation of protein activity *via* both *O*-GlcNAc and phosphate on the same protein would greatly increase the level of control that posttranslational modifications can exert on a given protein.

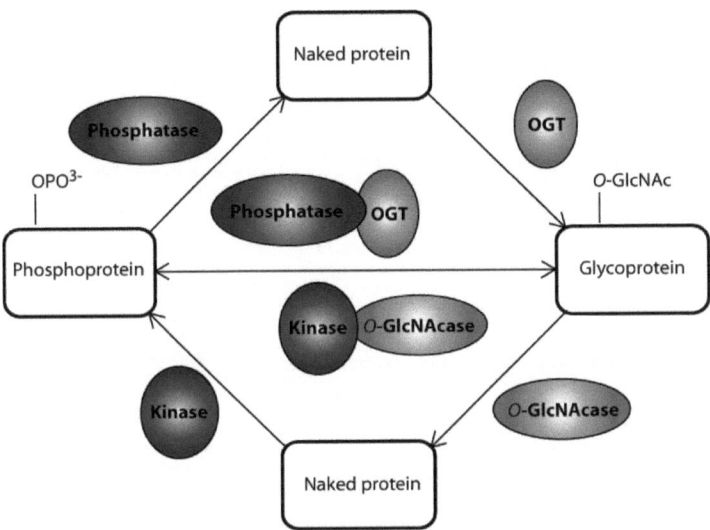

Fig. 13. The yin-yang hypothesis: *O*-GlcNAc prevents phosphorylation and *vice versa*. *O*-GlcNAcylation competes with phosphorylation for the same or adjacent sites. Therefore, proteins must be dephosphorylated before *O*-GlcNAcylation can occur and de-*O*-GlcNAcylated before phosphorylation can occur. Swapping between the modification state may regulate different functions of one protein. Figure modified from (252).

2.4 Aims of the Study

O-GlcNAc is a sensitive nutrient sensor coupling gene expression, signal transduction and protein degradation to the nutrition state of the cell and thus to the glucose metabolism. Stimulation of T lymphocytes with mitogens leads to alterations in cellular and nuclear *O*-GlcNAc levels. *O*-GlcNAc modification has further been linked to the regulation of gene transcription, as both enzyme involved in *O*-GlcNAc cycling, OGT and NCOAT, colocalize in a complex at the chromatin, together with other transcription-modulating proteins.

Some of the transcription factors involved in HIV-1 gene regulation are modified by *O*-GlcNAc, including AP-1, YY1, NFATc1, NF-κB and Sp1. Two decades ago, Sp1 was the first transcription factor identified to be modified by *O*-GlcNAc. Since then, many efforts have been made to characterize the effect of *O*-GlcNAcylation on Sp1 activity. Contradictory results have been published, describing on the one hand a decrease of Sp1 transactivation capability, on the other hand an increase in Sp1 activity upon its modification by *O*-GlcNAc.

Since Sp1 is a critical and important transcription factor involved in HIV-1 gene transcription, and Sp1 activity is modulated by *O*-GlcNAc, the aim of this study was to investigate the effect of Sp1 *O*-GlcNAcylation on the activity of the HIV-1 LTR promoter.

3. Material and Methods

3.1 Material

3.1.1 Chemicals, Media and Reagents

Table 2. Chemicals, media and reagents and their suppliers

Chemicals, media, reagent	Supplier
30% acrylamide / N,N – methylenebisacrylamide (29:1)	Bio-Rad
40% acrylamide / N,N – methylenebisacrylamide (29:1)	Bio-Rad
Agarose	Peqlab
Ammonium persulfate (APS)	Sigma
Ampicillin	Sigma
Anti-Sp1 Pep2 agarose conjugate	Santa Cruz
Bacto-agar	Difco
Blocking reagent	Roche
Boric acid	Sigma
Calcium chloride	Merck
Complete Mini, EDTA-free protease inhibitor cocktail	Roche
Desoxyribonucleotide triphosphate	Peqlab
Diethylene pyrocarbonate (DEPC)	Roth
Dimethyl sulfoxide	Sigma
Dithiothreitol (DTT)	Sigma
Dulbecco's Modified Eagle Medium (DMEM)	PAA
Ethylenediaminetetraacetic acid (EDTA)	Sigma
Enhanced chemoluminescence (ECL) detection system	Thermo Scientific
Ethanol	Merck
Ethidium bromide	Roth
Fetal calf sera (FCS)	Biochrom
First strand buffer (5×)	Invitrogen
Formaldehyde loading dye	Ambion
G418	PAA
Gentamicin	PAA
Glacial acetic acid	Merck
D-glucosamine hydrochloride	Sigma
Laemmli-buffer (2×)	Bio-Rad
L-glutamine	PAA
Human Serum	PAA
4-(2-hydroxyethyl)-1-piperazineethanesulfonic acid (HEPES)	Sigma
Igepal / Nonidet P40	Sigma

Table 2 - Continued

Chemicals, media, reagent	Supplier
Interleukin-2 (IL-2)	Roche
Isopropanol	Merck
Kanamycin	Gibco
Low fat milk powder	Saliter
Magnesium chloride	Sigma
Magnesium sulfate	Merck
β-mercaptoethanol (β-ME)	Roth
Methanol	Merck
Minimum essential medium (MEM) nonessential amino acids	PAA
Oligo $(dT)_{18}$ Primer	Fermentas
OptiMEM I reduced serum medium (without phenol red)	Invitrogen
Orange G	Sigma
γ-^{32}P-ATP (222 TBq/mMol)	Hartmann Analytic
PBS-powder (5×, 10×)	Biochrom
Penicillin/streptomycin	PAA
Phytohemagglutinin P (PHA-P)	Sigma
Polynucleotide kinase buffer	New England Biolabs
Poly(dI-dC)•poly(dI-dC)	GE Healthcare
RNAsin	Promega
RPMI 1640	PAA
Sepharose CL-6B	Sigma
Sodium azide (NaN_3)	Sigma
Sodium chloride (NaCl)	Merck
Sodium deoxycholate	Sigma
Sodium dodecyl sulfate (SDS)	Roth
Sodium hydrogen phosphate ($NaHPO_4$)	Sigma
Sodium pyruvate	PAA
N, N, N′, N′-tetramethylethylenediamine (TEMED)	Bio-Rad
Tris glycine transfer buffer (10×)	Bio-Rad
Tris glycine SDS running buffer (10×)	Bio-Rad
Tris(hydroxymethyl)-aminomethane (Tris)	Sigma
Triton-X-100	Sigma
Trypsin / EDTA	PAA
Tween20	Sigma
Very Low Endotoxin (VLE)-RPMI 1640 Medium	Biochrom

3.1.2 Consumables

Table 3. Consumables and their suppliers.

Consumables	Supplier
Cell culture flasks and plates	Nunc
4-mm cuvettes	Molecular Bioproducts
Eppendorf tubes	Eppendorf
Filter tips	Bio-Rad
Gel-blotting-paper	VWR
Hybond –P polyvinylidene fluoride (PVDF) membrane	Roth
Immobilon-P^{SQ} transfer membrane	Millipore
LumiNunc, 96-well plates	Nunc
Nuc Trap probe purification columns	Stratagene
Serological Pipettes	Greiner
Tris glycine gradient gels (4-20%)	Anamed
X-ray films Fuji RX	Fuji

3.1.3 Equipment

Table 4. Equipment required for this study and their manufacturer.

Equipment	Manufacturer
Centrifuge	Eppendorf / Beckmann
Developer	Agfa
Digital camera Canon Power Shot G3	Canon
EPS power supply 601	Amersham
FACSCalibur	BD Biosciences
Gel documentation	Bio-Rad
Gene Pulser Xcell	Bio-Rad
Horizontal gel electrophoresis system	Peqlab
Luminoskan Ascent	Thermo Fisher Scientific
Microplate reader model 680	Bio-Rad
Microscope	Zeiss
Mini-Vertical gel casters	Amersham
Nucleofector II	Amaxa
pH-Meter	Metrohm
Spectrophotometer GeneQuant	Amersham
Phosphoimager/Laser Scanner FLA5000	Fuji
Protein transfer system TE 22 Mini Tank Transfer Unit	Amersham
Table centrifuge	Eppendorf

Table 4 - Continued

Equipment	Manufacturer
Thermocycler	Biometra / Bio-Rad
Vertical gel electrophoresis system mighty small II SE 260	Amersham

3.1.4 Kits

Table 5. Kits used in this study and their manufacturer.

Kit	Manufacturer
Cell Line Nucleofector Kit V	Amaxa
CellTiter 96 Non-Radioactive Cell Proliferation Assay (MTT)	Promega
CytoTox 96 Non-Radioactive Cytotoxicity Assay (LDH)	Promega
DC protein assay	Bio-Rad
Luciferase reporter gene assay system	Promega
PureLink HiPure Maxi-DNA isolation kit	Invitrogen
Mini-DNA isolation kit	Peqlab
mMESSAGE mMACHINE T7 ultra kit	Ambion
Nuclear/cytosol fractionation kit	BioVision
PCR purification kit	Qiagen
QIAshredder columns	Qiagen
QuikChange XL site-directed mutagenesis kit	Stratagene
RNeasy Mini Kit	Qiagen

3.1.5 Enzymes

Table 6. Enzymes used in this study and their suppliers.

Enzyme	Supplier
7-methyl(3'-O-methyl)GpppG anti-reverse cap analog (ARCA)	Ambion
DNase I	Qiagen
Pfu-Ultra HS DNA polymerase	Invitrogen
Platinum Taq DNA polymerase	Invitrogen
Restriction enzymes	New England Biolabs
Shrimp alkaline phosphatase (SAP)	Roche
Superscript III Reverse Transcriptase	Invitrogen
T4-DNA ligase	Fermentas
T4-Polynucleotide Kinase	New England Biolabs
Turbo DNase	Ambion

3.1.6 Standards

Table 7. Standards used in this study and their suppliers.

Standard	Supplier
peqGOLD 100 bp DNA ladder	Peqlab
peqGOLD 1kb DNA ladder	Peqlab
Page Ruler prestained protein ladder	Fermentas
RNA ladder 0.25 – 9.5 kb	Invitrogen

3.1.7 Software

Table 8. Software used for the evaluation of the data presented in this study and their manufacturer.

Software	Manufactuere
Aida Image Analyzer v. 4.15	raytest GmbH
Adobe Photoshop CS2	Adobe Systems Incorporation
Adobe Illustrator CS2	Adobe Systems Incorporation
CellQuest Pro software	BD Biosciences
Corel Draw 12	Corel Corporation
EndNote X1	Thomson Reuters
GelDoc	Bio-Rad
HP scanjet	Hewlett-Packard Company
Microsofft Office	Microsoft
SPSS 16.0	SPSS Incorporation
Vector NTI	Invitrogen

3.1.8 Buffer and Solutions

Table 9. Buffers and solutions used in this study and their composition.

Buffer, solution	Concentration	Composition
BSB	100 mM 10 mM 10 mM 5%	Tris-HCl (pH 7.5) DTT EDTA glycerol
0.5% / 10% blocking solution	0.5% / 10%	blocking reagent in PBS
FACS buffer	5% 0.1%	FCS NaN_3 in PBS

Table 9 - Contiunued

Buffer, solution	Concentration	Composition
HBS (2×)	140 mM	NaCl
	1.5 mM	Na_2HPO_4
	50 mM	HEPES (pH 7.05)
IP lysis buffer	20 mM	Tris/HCl (pH 7.5)
	150 mM	NaCl
	5 mM	$MgCl_2$
	1%	IGEPAL
IP wash buffer	20 mM	Tris/HCl (pH 7.5)
	150 mM	NaCl
	5 mM	$MgCl_2$
	0.1%	IGEPAL
LB media	1%	Trypton
	0.5%	yeast extract
	0.5%	NaCl
	pH 7.0	adjusted with HCl
LB-agar	12%	bacto-agar in LB media
	100 µg/ml	ampicillin
	50 µg/ml	kanamycin
2.5% / 5% low fat milk solution		PBS/Tween
	2.5% / 5%	low fat milk powder
MOPS buffer (10×)	41.8 g	MOPS
	20 mM	sodium acetate
	10 mM	EDTA (pH 8.0)
	pH 7.0	adjusted with NaOH
	add 1 L	DEPC-ddH_2O
Orange-G dye	0.15%	Orange G
	40%	glycerol
PBS/Tween	0.1%	Tween 20
		in PBS
Protein loading dye (6×)	350 mM	Tris/HCl (pH 6.8)
	30%	glycerol
	10%	SDS
	0.6 M	DTT
	0.012%	bromophenolblue
SOC media	0.5 g	yeast extract
	2 g	trypton
	0.2 g	$MgCl_2$
	0.25 g	$MgSO_4$
	0.36 g	glucose
	add 100 ml	ddH_2O
STE buffer	100 mM	NaCl
	20 mM	Tris-HCl (pH 7.5)
	10 mM	EDTA
TAE (50×)	2 M	Tris/HCl
	100 mM	EDTA
	1 M	glacial acetic acid

Table 9 - Continuued

Buffer, solution	Concentration	Composition
TBE (10×)	0.89 M	Tris/HCl
	0.89 M	boric acid
	20 mM	EDTA (pH 8.0)
Transfer buffer for Immobilon membrane	100 ml	Tris glycine buffer (10×)
	700 ml	H_2O
	200 ml	methanol
Transfer buffer for PVDF membrane	100 ml	Tris glycine buffer (10×)
	870 ml	H_2O
	30 ml	methanol

BSB, Band shift buffer; DEPC, diethylene pyrocarbonate; DTT, dithiotreitol; EDTA, ethylenediamine tetraacetic acid; FACS, fluorescence activated cell sorting; FCS, fetal calf sera; HBS, HEPES buffered saline; HEPES, 4-(2-hydroxyethyl)-1-piperazineethanesulfonic acic; IP, immunoprecipitation; LB, lysogeny broth; MOPS, 4-Morpholinepropanesulfonic acid; PBS, phosphate buffered saline; SDS, sodium dodecyl sulfate; SOC, super optimal broth with catabolite repression; STE, sodium-tris-EDTA; TAE, tris-acetate-EDTA; TBE, tris-boric acid-EDTA.

3.1.9 Primary Antibodies

Table 10. List of the primary antibodies used in this study.

Primary antibody	Type	Species	Dilution	Origin
anti-human-GAPDH	monoclonal	mouse	1:70,000	Millipore
anti-GFP	monoclonal	mouse	1:10,000	Roche
anti-human actin	polyclonal	rabbit	1:1,000	Sigma
anti-human histone H1	monoclonal	mouse	1:1,000	Santa Cruz
anti-human-Lamin A/C (N18)	polyclonal	goat	1:1,000	Santa Cruz
anti-eukaryotic-O-GlcNAc (CTD 110.6)	monoclonal	mouse	1:1,000	Hiss Diagnostic
anti-human-OGT (TI14)	polyclonal	rabbit	1:1,000	Sigma
anti-human-Sp1 (Pep2)	polyclonal	rabbit	1:500	Santa Cruz
anti-Tat (1C9)	monoclonal	rat	1:150	E. Kremmer

3.1.10 Secondary Antibodies

Table 11. List of the secondary antibodies used in this study.

Secondary antibody	Species	Dilution	Origin
anti-mouse IgG-HRP	sheep	1:5,000	Dako
anti-rabbit IgG-HRP	donkey	1:5,000	Ge Healthcare
anti-rat IgG-HRP	goat	1:5,000	Dako
anti-goat-IgG-HRP	rabbit	1:5,000	Dako

Material and Methods

3.1.11 Oligonucleotides

3.1.11.1 siRNA-Oligonucleotides

Table 12. List of the oligonucleotides used for the RNA interference experiments.

siRNA	Target Sequence (5'→3')	Order Number	Origin
Hs_SP1_1	caggtgcaaaccaacagatta	SI00150976	Qiagen
Hs_OGT_7	aagattaatgttcttcataaa	SI02665131	Qiagen
Silencer GAPDH	unknown	AM4605	Ambion
Silencer Negative Control #1	unknown	AM4611	Ambion

3.1.11.2 Cloning Oligonucleotides

Table 13. List of the oligonucleotides used for cloning.

Name	Sequence (5'→3')	Application
5'-*Bam*HI-*Eco*RV-<u>Kozak</u>-ATG-ncOGT(1)	tag<u>ggatcc</u>gatatc<u>gccacc</u>atggcgtcttccgtgggcaacg	Cloning
3'-*Stu*I-ncOGT(718)	agatctat*caggcct*tgctcatag	Cloning
5'-*Mfe*I-LTR-fwd	ga*caattg*aagaaaaggggggactggaagggctaattcactc	Cloning
3'-*Xho*I-LTR-rev	ac*ctcgag*ttggctcactgcaacctctacctcctgggtgct	Cloning
5'-Mut(6)-Wobble-OGT	gaggcacggcaacctgtgcttagataagat**c**aa**c**gtgctg**c**a**c**aagccaccatatgaacatccaaaaga	Mutagenesis-PCR
3'-Mut(6)-Wobble-OGT	tcttttggatgttcatatggtgg**c**ttgtg**c**ag**c**acgttgatcttatctaagcacaggttgccgtgcctc	Mutagenesis-PCR
5'-wtLTR-mutSp1	ggggacttt**cc**aggg**a**ttcgtggcctgttcgggactgg**tt**agtggcga	Mutagenesis-PCR
3'-wtLTR-mutSp1	ctcgccact**aa**ccagtccc**gaa**caggccacg**aa**tccctggaaagtcccc	Mutagenesis-PCR

The Kozak-sequence is underlined and the transcription start codon (atg) is highlighted grey. Italic letters display restriction sites and letters in bold indicate nucleotide exchanges. The mutagenic primers were designed by the QuikChange primer design program. All oligonucleotides were purchased from Invitrogen.

3.1.11.3 Gel Shift Oligonucleotides

Table 14. List of the oligonucleotides used for the electrophoretic mobility shift assays.

Name	Target Sequence (5'→3')
sense wt LTR-Sp1	ggatc<u>gggagcgtggcctgggcgggact</u>g<u>gggagtggc</u>gagccc
antisense-wt LTR-Sp1	gggctcg<u>ccactccccagtcccgcccaggccacgctccc</u>gatcc
sense-mut LTR-Sp1	ggatc<u>ggg**a**ttcgtggcctgttcgggactgg**tt**agtggc</u>gagccc
antisense-mut LTR-Sp1	gggctcgccact**aa**ccagtccc**gaa**caggccacg**aa**tcccgatcc

Underlined letters indicate the Sp1-binding sites; bold nucleotides display nucleotide exchanges compared to the wild-type sequence. wt = wilt-type, mut = mutated. All oligonucleotides were purchased from Invitrogen.

3.1.11.4 Oligonucleotides for RT-PCR

Table 15. List of the oligonucleotides used for the electrophoretic mobility shift assays.

Name	Target Sequence (5'→3')
Jur-OGT-fwd_1-3	tccagtgtggtggaattctg
Jur-OGT-rev_1	ttgcgtctcaattgctttca
GAPDH-fwd	agccacatcgctcagaacac
GAPDH-rev	gaggcattgctgatgatcttg

All oligonucleotides were purchased from Invitrogen.

3.1.11.5 Sequencing Oligonucleotides

Table 16: Oligonucleotides used for sequencing of the cloned constructs.

Name	Complementary to	Sequence (5'→3')
5'-*Eco*RV-ncOGT(aa1)	pcDNA4-ncOGT	tag*gatatc*atggcgtcttccgtgggcaacg
5'-ncOGT(501)-fwd	pcDNA4-ncOGT	ctctgctcttcagtacaatcctg
5'-ncOGT(1078)-fwd	pECE-mOGT; pcDNA4-ncOGT	gtcttcccagagtttgctgctgcc
5'-ncOGT(1721)-fwd	pECE-mOGT; pcDNA4-ncOGT	tcagtgatggtcggctgcgtg
5'-ncOGT(2342)-fwd	pECE-mOGT; pcDNA4-ncOGT	gtcctgatggaggagacaatgc
5'-mOGT(2960)-fwd	pECE-mOGT; pcDNA4-ncOGT	gagcattatgcagctggcaac
3'-*Not*I-ncOGT(aa1046)-rev	pECE-mOGT; pcDNA4-ncOGT	tag*cgccggcg*ttatgctgactcagtgacttcaacagg
3'-ncOGT(2601)rev	pECE-mOGT; pcDNA4-ncOGT	acgcaacagccagagtacactattg
3'-ncOGT(1900)-rev	pECE-mOGT; pcDNA4-ncOGT	gcaaaaagctcatttcgagcgccc
3'-ncOGT(1160)-rev	pECE-mOGT; pcDNA4-ncOGT	tgggcagcagcaaactctggg
3'-ncOGT(464)-rev	pcDNA4-ncOGT	gctaccaaggcggctgccagg
pcDNA4-d2EGFP(5001)-fwd	pcDNA4	caaatagggttccgcgcac
pcDNA4-LTR-d2EGFP(5601)-fwd	HIV-1 LTR	gaacccactgcttaagcctc
pcDNA4-d2EGFP(552)-fwd	d2EGFP	gccgacaagcagaagaacg
pcDNA4-d2EGFP(1058)-rev	pcDNA4	gcaactagaaggcacagtcg
pcDNA4-d2EGFP(552)-rev	d2EGFP	cgttcttctgcttgtcggc
pcDNA4-LTR-d2EGFP(5682)-rev	HIV-1 LTR	gggtctgagggatctctag
pcDNA4- EF1α-d2EGFP(5645)-fwd	EF1α	ggccaagatctgcacactgg
pcDNA4- EF1α-d2EGFP(6201)-rev	EF1α	ctgaggcttgagaatgaacc
pcDNA4- EF1α-d2EGFP(5645)-rev	EF1α	ccagtgtgcagatcttggcc
Sp1(525)-fwd	pcDNA3-Sp1	gttccagaccgttgatggg
Sp1(1197)-fwd	pcDNA3-Sp1	gcctcagctagttcaagggggg
Sp1(1874)-fwd	pcDNA3-Sp1	gccacatccaaggctgtggg
Sp1(1807)-rev	pcDNA3-Sp1	ggtgcatgcttcccgccggg
Sp1(1238)-rev	pcDNA3-Sp1	cctggagggcctgtccccttg
Sp1(586)-rev	pcDNA3-Sp1	ctgctgcacttgggccccag

Table 16 - Continued

Name	Complementary to	Sequence (5'→3')
Luc-rev	*firefly*-Luciferase	agttgctctccagcggttcc
M13-fwd	pCR2.1	gtaaaacgacggccag
M13-rev	pCR2.1	caggaaacagctatgac
Bgh-rev	pcDNA4	tagaaggcacagtcgagg
T7-fwd	pcDNA4	taatacgactcactataggg

The transcription start and stop codons are highlighted grey and italic letters indicate restriction sites. The numbers in brackets indicate the position of the first bound nucleotide or amino acid (aa). fwd = sense strand, rev = antisense strand. All oligonucleotides were purchased from Invitrogen.

3.1.12 Plasmids

Table 17. Plasmids used in this study.

Name	Insert	Description	Reference
pcDNA4/*myc*-His B, pcDNA3.1(-)		Empty vector	Invitrogen
pGL3-Basic		Promoterless vector carrying luciferase	Promega
pEF1/*myc*-His A		Empty vector containing the EF1α promoter	Invitrogen
pGL3-EF1α-Luc	Luciferase	Expresses luciferase under the control of the pEF1α promoter	This work
pECE-OGT/Lv4F	mOGT	Expresses the Lv4F fragment, mitochondrial OGT	(147)
pcDNA4-mOGT	mOGT	Expresses the Lv4F fragment, mitochondrial OGT	This work
pcDNA4-OGT	wtOGT	Expresses the nucleocytoplasmatic OGT	This work
pcDNA4-resOGT	resOGT	Expresses resOGT, an rescue mutant with six silent nucleotide exchanges	This work
pIL2-d2EGFP	IL2d2EGFP	Expresses d2EGFP under the control of the IL-2 promoter	Florian. Forster, Vienna
pBS-Sp1-fl	Sp1	Phagemic vector containing the Sp1 cDNA	(109)
pcDNA3(-)-Sp1	Sp1	expresses Sp1	This work
pCEP4-Tat	Tat	Expresses HIV-1 Tat	(22); AIDS Reagent Program;
pNL4-3	HIV-1 genome	Plasmid encoding the HIV-1 genome	(2); AIDS Reagent Program,
pNL4-3.Luc.R⁻E⁻	HIV-1 genome; Luciferase in the nef gene	Reporter construct expressing luciferase under the control of the wild-type HIV-1 LTR promoter	(34, 90); AIDS Reagent Program,
pXP1-LTR-κB-Sp1$_{wt}$-Luc	Luciferase	Reporter construct expressing luciferase under the control of the HIV-1 LTR consisting of the NF-κB and Sp1-binding sites	(9, 70)

Table 17 - Continued

Name	Insert	Description	Reference
pcDNA4-d2EGFP	CMV-d2EGFP	Reporter construct expressing d2EGFP under the control of the CMV promoter	This work
pcDNA4-d2EGFP ∅ CMV promoter	d2EGFP	Promoterless vector carrying d2EGFP	This work
pcDNA4-LTRwt-d2EGFP	LTRwt-d2EGFP	Reporter construct expressing d2EGFP under the control of the HIV-1 LTRwt promoter	This work
pcDNA4-LTRmutSp1-d2EGFP	LTRmutSp1-d2EGFP	Reporter construct expressing d2EGFP under the control of the HIV-1 LTRmutSp1 promoter	This work
pcDNA4-EF1α-d2EGFP	EF1α-d2EGFP	Reporter construct expressing d2EGFP under the control of the EF1α promoter	This work
pXP1- LTR-κB-Sp1$_{mut}$-Luc	Luciferase	Reporter construct expressing luciferase under the control of the HIV-1 LTR consisting of the NF-κB-binding sites. The Sp1-binding sites are mutated	(9, 70)
pXP1- LTR-Sp1$_{wt}$-Luc	Luciferase	Reporter construct expressing luciferase under the control of the HIV-1 LTR consisting of the Sp1-binding sites.	(9, 70)
pXP1- LTR-Sp1$_{mut-Sp-III}$-Luc	Luciferase	Reporter construct expressing luciferase under the control of the HIV-1 LTR consisting of the Sp1-binding motifs with mutation in the Sp1-binding site III (Sp-III)	(9, 70)
pXP1- LTR-Sp1$_{mut-Sp-II}$-Luc	Luciferase	Reporter construct expressing luciferase under the control of the HIV-1 LTR consisting of the Sp1-binding motifs with mutation in the Sp1-binding site II (Sp-II)	(9, 70)
pXP1- LTR-Sp1$_{mut-Sp-I}$-Luc	Luciferase	Reporter construct expressing luciferase under the control of the HIV-1 LTR consisting of the Sp1-binding motifs with mutation in the Sp1-binding site I (Sp-I)	(9, 70)
pXP1- LTR-Sp1$_{mut-Sp-I+II}$-Luc	Luciferase	Reporter construct expressing luciferase under the control of the HIV-1 LTR consisting of the Sp1-binding motifs with mutations in the Sp1-binding site I and II (Sp-I+II)	(9, 70)
pGEM4Z-5'UTR-sig-huSurvivin-DC.LAMP-3'UTR-64A	Human Survivin	RNA-production vector for human Survivin mRNA	(12)
pGEM4Z-EGFP-64A	EGFP	RNA-production vector for EGFP mRNA	(199)
pGEM4Z-OGT-64A	OGT	RNA-production vector for OGT mRNA	This work

Material and Methods

The plasmids constructed in this work were cloned as followed:

pGL3-EF1α-Luc :   EF1α promoter was cut out from pEF1alpha/*myc*-HisA with *Hind*III and inserted into the promoterless pGL3-Basic vector.

pcDNA4-mOGT: C-terminal fragment cut out from Lv4F (U77413) *via EcoR*V and *Not*I and cloned into pcDNA4-mycHisB. The missing N-terminus (38 nts) was filled up with oligonucleotides *via Kpn*I and *EcoR*V.

pcDNA4-OGT: PCR-amplification of the N-terminal 820 nts with human cDNA as template and subcloning into the pCR2.1 vector. C-terminus cut out from pECE-OGT/Lv4F and subcloned together with N-terminus-ncOGT into pcDNA4/*myc*-His B *via EcoR*V and *Not*I.

pcDNA4-resOGT: Generated *via* mutagenesis PCR using pcDNA4-OGT as template and the primer 5'-Mut(6)-Wobble-OGT and 3'-Mut(6)-Wobble-OGT.

pcDNA3(-)-Sp1:   Sp1 was cut out with *Xba*I and *Sma*I from pBS-Sp1-fl and subcloned in pcDNA3.1(-) linearized with *Xba*I and *EcoR*V. cDNA encodes AF252284.

pcDNA4-d2EGFP: d2EGFP cut out from pIL2-d2EGFP with *Xho*I and *Xba*I and subcloned into pcDNA4/*myc*-His B.

pcDNA4-d2EGFP ∅ CMV promoter:
     the CMV promoter was cut out from pcDNA4-d2EGFP with *Bgl*II and *EcoR*V. The remaining backbone was filled up with Klenow and ligated.

pcDNA4-LTRwt-d2EGFP: PCR of the 3'wt-LTR using pNL4-3 as template and the primers 5'-*Mfe*I-LTR-fwd and 3'-*Xho*I-LTR-rev. The PCR fragment was subcloned into the pcDNA4-d2EGFP vector *via Mfe*I and *Xho*I.

pcDNA4-LTRmutSp1-d2EGFP:
     Mutagenesis PCR using pcDNA4-LTRwt-d2EGFP as template and the mutagenesis primers 5'-wtLTR-mutSp1 and 3'-wtLTR-mutSp1

pcDNA4-EF1α-d2EGFP: EF1alpha-Promoter cut out from pGL3-EF1alpha-Luc *via Hind*III and inserted into the *Xho*I-linearized pcDNA4-d2EGFP ∅ CMV promoter.

pGEM4Z-OGT-64A: Digest of pGEM4Z-5'UTR-sig-huSurvivin-DC.LAMP-3'UTR-64A with *Xho*I. Klenow reaction was performed to fill up the 5' overhang. Afterwards, digest with *Bgl*II to remove the huSurvivin-DC.LAMP.

The pcDNA4-OGT vector was digested with *Not*I. Klenow reaction was performed to fill up the 5' overhang. Subsequently digested with *Bam*HI to obtain ncOGT. Ligation *via* the compatible ends BglII/BamHI and blunt end.

3.1.13 Biological Material

3.1.13.1 Bacterial Strains

XL1-Blue (Stratagene): *recA1 endA1 gyrA96 thi-1 hsdR17 supE44 relA1 lac* [F' *proAB lacI^q ZΔM15* Tn*10* (Tet^r)].

XL10-Gold (Stratagene): Tet^{r}Δ *(mcrA)183 Δ(mcrCB-hsdSMR-mrr)173 endA1 supE44 thi-1 recA1 gyrA96 relA1 lac* Hte [F' *proAB lacI^q ZΔM15* Tn*10* (Tet^r) Amy Cam^r].

3.1.13.2 Eukaryontic Cell Lines

HEK 293T: human embryonic kidney cells, stably transformed with the human Adenovirus Type 5; the cells further contain the SV40 large T Antigen (47).

Jurkat: human T cell leukemia cell line established from the peripheral blood of a 14 year old boy (200).

T1 (174 × CEM.T1): hybrid cell line between the B lymphoblastic 721.174 and the T lymphoblastic CEMR.3 cell lines (195).

HeLa-Tat-III/LTR/d1EGFP: HPV18-positive human epithelial cervical cancer cell line, stably expressing the enhanced green fluorescent protein (EGFP)under the control of the HIV-1 LTR promoter. The cells were purchased from the NIH AIDS Research and Reference Reagent Program (171).

3.1.13.3 Primary Cells

CD4⁺ T lymphocytes: were isolated from blood using anti-CD4 MACS beads (Miltenyi) in cooperation with the Department of Dermatology of the University of Erlangen.

3.1.13.4 Recombinant Viruses

$HIV_{NL4-3.Luc.R-E-}$: contains the *firefly* luciferase gene in the *nef* open reading frame. The two frame shift mutations in the *env*- and *vpr*-genes require pseudotyping with the vesicular stomatitis virus (VSV)-G envelope. Hence, this pseudovirus can perform only one round of replication.

3.2 Methods

3.2.1 RNA-Techniques

3.2.1.1 Isolation of Cellular RNA from Eukaryotic Cells

Isolation of cellular RNA from Jurkat cells was performed *via* the RNeasy Mini-Kit according to the manufacturer's instructions. The cells (7×10^6) were washed once with PBS and lysed in 600 µl RLT buffer (provided with the kit) complemented with β-ME. The lysate was homogenized with QIAshredder spin columns to shear the chromosomal DNA. To eliminate genomic DNA contamination, on-column DNase I digest was performed. The cellular RNA was eluted in 50 - 55 µl RNase-free water (DEPC-ddH$_2$O). The RNA concentration was measured spectrophotometrically and the quality of the RNA was verified by agarose gel electrophoresis.

3.2.1.2 Agarose Gel Electrophoresis of RNA

RNA was electrophoretically separated in 1% (w/v) agarose gels containing ethidium bromide (5µg/ml) in 1× MOPS buffer. The RNA samples were mixed with Orange-G dye (diluted in DEPC-ddH$_2$O) and loaded onto the gels. The RNA ladder 0.25 – 9.5 kb was used as a size standard. Gels were run in 1× MOPS buffer at approximately 100 Volt (V) and documented using the GelDoc system.

3.2.1.3 Quantitative Measurement of RNA

The amount of RNA was quantified with a spectrophotometer at a wavelength of 260 nm in a quarz cuvette with a coat thickness of 1 cm. The extinction value ($A_{260\,nm}$) = 1.0 corresponds to 40 µg/ml RNA. RNA concentration was calculated according to the following formula: amount [µg/ml] = optical density at 260 nm × dilution factor × 40 µg/ml (amount of RNA with optical density at 260 nm = 1). Since proteins absorb light at a wavelength of 280 nm, RNA purity can be determined by measuring at both wavelengths. The quotient of the optical density at 260 nm and 280 nm must be over 2.0 in order to regard the measured RNA as pure. RNA preparations performed in this study had a quotient higher than 2.0.

3.2.1.4 In vitro Transcription of mRNA

In vitro transcription of RNA was performed as described previously (198) using the mMESSAGE mMACHINE T7 Ultra kit. The pGEM4Z RNA-synthesis vector encoding either OGT or EGFP were used as templates. The DNA was linearized by *Not*I or *Spe*I, respectively and purified by phenol/chloroform/ethanol precipitation. The *in vitro* transcription was performed with T7 RNA polymerase according to the manufacturer's instructions in a total volume of 60 µl:

a) ARCA *in vitro* transcription for 60 µg scale:

    3 µg template DNA
    30 µl 2× NTP/ARCA
    6 µl T7 Reaction Buffer (10×)
    6 µl T7 Enzyme Mix
    16.5 µl nuclease-free H_2O
    → Incubation for 3 h at 37°C
    add 3 µl Turbo DNase
    → Incubation for 1.5 h at 37°C

b) Poly A tailing of the *in vitro* transcribed RNA:

    108 µl nuclease-free H_2O
    60 µl E-PAP buffer (5×)
    30 µl $MnCl_2$ (25 mM)
    30 µl ATP Solution
    12 µl E-PAP Enzyme
    → Incubation for 1 h at 37°C

c) Purification of the *in vitro* transcribed RNA on RNeasy columns:

    450 µl RLT buffer complemented with β-ME were mixed with 330 µl ethanol;
    RNA was transferred to the RLT-ethanol mix and loaded onto RNeasy spin columns;
    columns were washed according to the manufacturer's instructions;
    RNA was eluted in 40 µl nuclease-free H_2O;
    The RNA was stored in aliquots at -80°C.

RNA quality was verified by agarose gel electrophoresis and RNA concentration was measured spectrophotometrically.

3.2.2 cDNA Synthesis

The protein-encoding mRNA was reversely transcribed in complementary single-stranded DNA (cDNA). Each reaction contained 1 µg of isolated cellular RNA, 10 mM DTT, 40 Units RNasin, 0.5 mM dNTP mix, 0.1 µM Oligo-dT(18)-primer (5´-TTTTTTTTTTTTTTTTTT-3´), 5× first strand buffer and 200 Units of the enzyme Superscript III Reverse Transcriptase. The Oligo-dT(18)-primer binds specifically to the poy-A tail of the mRNA and leads thereby to a selective transcription of the mRNA to cDNA. The cDNA synthesis was carried out with the Thermocycler under the following conditions:

        1µg total RNA
        1µl Oligo-dT(18)-primer
        1µl dNTP mix (10mM each)
        ad 12 µl DEPC-ddH$_2$O

a) Denaturation for 5 min at 65°C

        Add 4µl 5× first strand buffer
        2µl DTT
        1µl RNAsin
        → Incubation for 2 min at 42°C

b) First strand synthesis

        Add 1µl RT (SuperscriptIII)
        or 1µl DPEC H$_2$O for negative control
        →50min @ 42°C
          10min @ 50°C
          10min @ 60°C
          15min @ 70°C
        → the single-stranded cDNA was stored at -20°C

3.2.3 DNA-Techniques

3.2.3.1 Polymerase Chain Reaction (PCR)

a) Reverse Transcription (RT)-PCR

cDNA synthesized from Jurkat cells stably transfected with d2EGFP or OGT was used as template. The RT-PCR was carried out with either undiluted or diluted (1:5, 1:10) cDNA. The reactions were carried out in a final volume of 25 µl with the following components:

2.5µl 10× PCR buffer
0.75ml MgCl$_2$
0.2µl dNTP mix (25mM each)
0.125µl fwd primer (100pmol)
0.125µl rev primer (100pmol)
0.125µl Platinum Taq DNA polymerase
1µl cDNA

Primers specifically amplifying recombinant OGT were selected: the forward primer recognized specifically the vector-encoded 5' untranslated region of the recombinant OGT cDNA, while the reverse primer was homologous to a sequence corresponding to the N-terminal region of both, endogenous and recombinant OGT. The PCR was carried out with the Thermocycler under the following conditions:

1. Initial denaturation	5 min	94°C	
2. Denaturation	30 s	94°C	
3. Hybridization	30 s	60°C	
4. Synthesis	60 s	72°C	35 cycles
5. Repetition of the steps 2 - 4			
6. Final synthesis	7 min	72°C	
7. Chill	∞	4°C	

Amplification of GAPDH was performed using the GAPDH-fwd and GAPDH-rev primers as described previously (141). The PCR was carried out with the Thermocycler under the following conditions:

1. Initial denaturation	5 min	94°C	
2. Denaturation	30 s	94°C	
3. Hybridization	30 s	51°C	
4. Synthesis	60 s	72°C	25 cycles
5. Repetition of the steps 2 - 4			
6. Final synthesis	7 min	72°C	
7. Chill	∞	4°C	

b) Cloning-PCR

Amplifications of DNA fragments were performed with a mixture of 16:1 Units Platinum-Taq DNA polymerase and Pfu-Ultra HS DNA polymerase in the corresponding reagents according to the manufacturer's protocol.

cDNA synthesized from HEK 293T cells was used as template. The reactions were carried out in a final volume of 50 µl with the following components:

1 µl cDNA
5 µl PCR buffer (10×)
5 µl dNTP mix (2.5 mM)
5 µl fwd primer (10 pmol)
5 µl rev primer (10 pmol)
0.5 µl DNA polymerase mixture

The PCR was carried out with the Thermocycler under the following conditions:

1. Initial denaturation   2 min   94°C
2. Denaturation   35 s   94°C
3. Hybridization   40 s   52°C
4. Synthesis   2 min   72°C   25 cycles
5. Repetition of the steps 2 - 4
6. Final synthesis   10 min   72°C
7. Chill   ∞   4°C

The PCR was performed with the following oligonucleotides in order to obtain the N-terminal sequence of OGT (1 – 818): 5'-*Bam*HI-*Eco*RV-Kozak-ATG-ncOGT(1) and 3'-*Stu*I-ncOGT(718). The forword primer contained sites for the restriction enzymes *Bam*HI and *Eco*RV (italic), Kozak sequence (underlined) and the transcription start codon ATG. The reverse primer contained the site for the restriction enzyme *Stu*I (italic). The amplified fragment was verified by gel electrophoresis on a 1.2% agarose gel and subsequently ligated into the cloning vector pCR2.1.

c) Mutagenesis-PCR

In order to generate a plasmid encoding an OGT rescue mutant (pcDNA4-resOGT), which escapes silencing by the wtOGT specific siRNA, six silent nucleotide exchanges were introduced in the siRNA binding region of wtOGT. The plasmid pcDNA4-LTRmutSp1-d2EGFP, which expresses d2EGFP under the control of the HIV-1 LTRmutSp1, was generated by introducing nucleotide exchanges in the HIV-1 wtLTR promoter. The nucleotide exchanges were created by mutagenesis PCR using as template pcDNA4-OGT and pcDNA4-LTRwt-d2EGFP, respectively and the QuikChange XL Site-Directed Mutagenesis Kit according to the manufacturer's instructions. The reactions were carried out with in a final volume of 50 µl with the following components:

10 ng template
5 µl PCR buffer (10×)
1 µl dNTP mix
125 ng fwd primer
125 ng rev primer
3 µl Quik Solution
1 µl Pfu Turbo DNA polymerase

The mutagenic primers 5'-Mut(6)-Wobble-OGT, 3'-Mut(6)-Wobble-OGT, 5'-wtLTR-mutSp1 and 3'-wtLTR-mutSp1 were designed by the QuikChange Primer Design Program. The reactions were carried out with the Thermocycler under the following conditions:

1. Initial denaturation	1 min	95°C	
2. Denaturation	50 s	95°C	
3. Hybridization	50 s	60°C	
4. Synthesis	2 min / kb	68°C	19 cycles
5. Repetition of the steps 2 - 4			
6. Final synthesis	7 min	68°C	
7. Chill	∞	4°C	

3.2.3.2 Agarose Gel Electrophoresis of DNA

DNA was electrophoretically separated in 1-1.5% (w/v) agarose gels containing ethidium bromide (5µg/ml) in 1× TAE buffer. The DNA samples were mixed with Orange-G dye and loaded onto the gels. The peqGOLD 1kb DNA-ladder or peqGOLD 100 bp DNA-ladder were used as DNA size standards. Gels were run in 1× TAE buffer at approximately 100 Volt (V) and documented using the GelDoc system (Biorad).

3.2.3.3 Isolation of DNA Fragments from Gels

Elektrophoretic separated DNA fragments were visualized with UV light and cut out with a scalpel. DNA fragments were isolated with the QIAquick Gel Extraction Kit according to the manufacturer's instructions. The DNA was eluted in 30 µl ddH$_2$O.

3.2.3.4 Restriction Digest of DNA Molecules

DNA was digested with restriction enzymes from New England Biolabs or Fermentas, in the appropriate buffer. The reactions were performed in a total volume of 10 µl for characterization and identification of DNA molecules, 100 µl for cloning, or 200 µl for template linearization used for the *in vitro* transcription of mRNA. The reactions were incubated at the temperature recommended by the manufacturer for 1h or over night.

Final volume 10 µl:	Final volume 100 µl:	Final volume 200 µl:
0.2 – 0.5 µg DNA	2 – 5 µg DNA	100 µg DNA
1 µl restriction buffer (10x)	10 µl restriction buffer (10x)	20 µl restriction buffer (10x)
5 U restriction enzyme	50 U restriction enzyme	100 U restriction enzyme
ev. 1 µl BSA (10x)	ev. 1 µl BSA (100x)	2 µl BSA (100x)

Table 18. List of the restriction enzymes used in this study.

Restriction enzyme	Recognition site	Emerging end
AgeI	A/CCGGT	5'-overhang
BamHI	G/GATCC	5'-overhang
BglII	A/GATCT	5'-overhang
DpnI	GA$_{CH3}$/TC	blunt
EcoRV	GAT/ATC	blunt
HindIII	A/AGCTT	5'-overhang
KpnI	GGTAC/C	3'-overhang
MfeI	C/AATTG	5'-overhang
NcoI	C/CATGG	3'-overhang
NdeI	CA/TATG	5'-overhang
PmeI	GTTT/AAAC	blunt
PstI	CTGCA/G	3'-overhang
PvuI	CGAT/CG	3'-overhang
PvuII	CAG/CTG	blunt
ScaI	AGT/ACT	blunt
SmaI	CCC/GGG	blunt
SpeI	A/CTAGT	5'-overhang
StuI	AGG/CCT	blunt
NotI	GC/GGCCGC	5'-overhang
XbaI	T/CTAGA	5'-overhang
XhoI	C/TCGAG	5'-overhang

3.2.3.5 Purification of Linearized DNA with Phenol/Chloroform/Ethanol

One volume of phenol:chloroform:isoamylalcohol (25:24:1) was added to the restriction digest and votrexed for1 min. The eppi was vigorously shaken and the mixture was centrifuged for 2 min at top speed and 4°C. The upper phase was collected and mixed with 1 volume of chloroform:isoamylalcohol (24:1). After vortexing for 1 min, the eppi was vigorously shaken and the mixture was centrifuged for 2 min at top speed and 4°C. The upper phase was collected again. 0.1 volume of 3 M calcium acetate (pH 5.6) was added together with 2.5 volumes of ethanol abs. The eppi was vortexed for 1 min and vigorously shaken before incubation for 1 h at -20°C to precipitate the DNA. Subsequently, the mixture was centrifuged at top speed for 30 min at 4°C, the supernatant was discarded and the DNA pellet was washed 2× with 500 μl 70% ethanol. Each centrifugation step was performed for 15 min at 4°C. Finally, the DNA pellet was washed once with 500 μl ethanol abs., centrifuged for 5 min at top speed and 4°C, and dried under the laminar flow. The DNA pellet was diluted in 30 μl DEPC-ddH$_2$O and the concentration of the purified DNA was determined.

3.2.3.6 Dephosphorylation of Vector DNA

The linearized vector DNA was dephosphorylated at the 5´ends with shrimp alkaline phosphatase (SAP) to inhibit backbone religation without insert during a ligation reaction. The reactions were carried out with in a final volume of 30 µl with the following components:

> 2.5 µg linearized vector DNA
> 3 µl SAP
> 3 µl SAP buffer

The reactions were incubated for 1 h at 37°C. The SAP enzyme was subsequently heatinactivated for 20 min at 65°C.

3.2.3.7 Klenow Reaction

To enable the ligation of two molecules, whose ends do not fit together, 5' overhangs were filled in with the DNA Polymerase I, Large (Klenow) Fragment to form blunt ends. DNA Polymerase I, Large (Klenow) Fragment is a proteolytic product of *E. coli* DNA Polymerase I which retains polymerization and 3'→ 5' exonuclease activity, but has lost 5'→ 3' exonuclease activity. Klenow retains the polymerization fidelity of the holoenzyme without degrading 5' termini. The reactions were carried out with 1 U/µg DNA DNA Polymerase I, Large (Klenow) Fragment and 200 µM dNTP's (each) in NEB buffer 2 or 4.

3.2.3.8 Ligation of DNA

Ligation of DNA fragments was performed using the Rapid DNA Ligation kit as suggested by the manufacturer using 5 Units T4-DNA-Ligase (Fermentas) and the corresponding T4 buffer. 100 ng vector DNA were ligated with a 6-fold molar excess of insert DNA. The ligation was carried out in a total volume of 20 µl for at least 20 min at RT.

3.2.3.9 Heat Shock Transformation of Bacteria

For transformation of bacteria with plasmid DNA or ligation reactions, 30 µl of competent *E. coli* XL-1 blue cells were pre-incubated on ice for 30 min with 50 – 100 ng plasmid DNA or 2 µl ligation reaction, respectively. The cells were then heat shocked for exactly 45 sec at 42°C in a water bath followed by 2 min incubation on ice. Transformed cells were subsequently incubated in 500 µl pre-warmed SOC medium for 1 h at 200 revolutions per minute (rpm) and plated on LB agar plates. After incubation of the agar plates at 37°C over night, single colonies were inoculated in LB media and analyzed for positive clones.

3.2.3.10 Isolation of Plasmid DNA

A 200 ml overnight culture of *E.coli* bacteria grown in LB-media containing antibiotics was centrifuged at 4,000 rpm for 10 min at 4°C. After decanting the supernatant, the pellet was employed in a plasmid maxi preparation using the PureLink HiPure plasmid maxiprep kit. The manufacturer's standard protocol was modified by using endotoxin removal wash buffer containing 100 mM sodium acetate (adjusted to pH 5.0 with acetic acid), 750 mM sodium chloride and 1% (w/v) Triton X-100. Finally, plasmid pellets were dried and then resuspended in 100 - 200 µl TE buffer supplied by the manufacturer. For preparation of 20 – 50 µg plasmid DNA, 5 ml overnight culture of *E.coli* bacteria grown in LB-media containing antibiotics was centrifuged at 10,000 rpm for 1 min at RT. After decanting the supernatant, the pellet was employed in a plasmid mini preparation using the Mini-DNA isolation kit according to the manufacturer's instructions.

3.2.3.11 DNA Sequencing

All fragments amplified by PCR and subsequently cloned into eukaryotic vectors were fully sequenced. Sequencing was performed using the method of Sanger et al. (197) and the Big Dye Terminator Mix (Perkin Elmer), which contains a thermostable DNA polymerase, unlabeled dNTP's and fluorescence-labeled ddNTP's, whose fluorescence is amplified due to the Big Dye. Sequencing reactions were carried out in a total volume of 10 µl and contained the following components:

> 150 - 200 ng DNA
> 7.5 pmol sequencing primer
> 1.75 µl TM buffer (5×)
> 0.5 µl Big Dye

The asymmetrical PCR for the automated sequencing was performed in a Thermocycler under the following conditions:

1. Initial denaturation	3 min	96°C	
2. Denaturation	10 s	96°C	
3. Hybridization	10 s	50°C	
4. Synthesis	4 min	60°C	24 cycles
5. Repetition of the steps 2 – 4			
6. Chill	∞	4°C	

Ethanol precipitation and sequencing was performed by the Institute of Human Genetics, University of Erlangen.

3.2.3.12 Quantitative Measurement of DNA

The amount of DNA was quantified with a spectrophotometer at a wavelength of 260 nm in a quarz cuvette with a coat thickness of 1 cm. The extinction value ($A_{260\ nm}$) = 1.0 corresponds to 50 µg/ml double-stranded DNA. DNA concentration was calculated according to the following formula: amount [µg/ml] = optical density at 260 nm × dilution factor × 50 µg/ml (amount of plasmid DNA with optical density at 260 nm = 1). Since proteins absorb light at a wavelength of 280 nm, the purity of the DNA can be determined by measuring at both wavelengths. The quotient of the optical density at 260 nm and 280 nm must be between 1.8 and 2.0 in order to regard the measured DNA as pure. For plasmid preparations used in this study, quotients were higher than 1.8.

3.2.4 Cell Biological Methods

3.2.4.1 Cultivation of E.Coli

Over night cultures of E.coli bacterial cells were maintained in LB media containing antibiotics (either 100 µg/ml ampicillin or 50 µg/ml kanamycin), incubated at 37°C and shaked in an incubator at 200 rpm. For long term storage, glycerol stocks containing 800 µl bacteria culture and 200 µl glycerol were stored at -80°C.

3.2.4.2 Cultivation of Mammalian Cells

Jurkat and T1 cells were cultured in VLE-RPMI 1640 supplemented with 10% FCS and 2 mM L-glutamine at 37°C and 5% CO_2. Cells were splitted every 2-3 days at a ratio 1:6.

Stably transfected Jurkat cells were cultured in VLE-RPMI 1640 supplemented with 10% FCS, 2 mM L-glutamine and 400 µg/ml Zeocin at 37°C and 5% CO_2. Cells were splitted every 2-3 days at a ratio 1:6.

HeLa-Tat-III/LTR/d1EGFP cells were cultured in DMEM supplemented with 10% FCS, 2 mM L-glutamine and 1 mg/ml G418 at 37°C and 8.5% CO_2.

HEK 293T cells were cultured in DMEM supplemented with 10% FCS and 2 mM L-glutamine at 37°C and 8,5% CO_2.

Stably transfected HEK 293T cells were cultured in DMEM supplemented with 10% FCS, 2 mM L-glutamine and 600 µg/ml Zeocin at 37°C and 8.5% CO_2.

CD4$^+$ **primary T lymphocytes** were cultured in MLPC media consisting of RPMI 1640 supplemented with 10% human serum, 10 mM HEPES, 1 mM sodium pyruvate, 2 mM L-glutamine, 500 µg/ml gentamicin and 1× non-essential amino acids at 37°C and 5% CO_2.

3.2.4.3 HIV-1 Pseudovirus Production

Pseudoviruses were produced by cotransfection of HEK 293T cells with pNL4-3.Luc.R-E- [obtained through the AIDS Research and Reference Reagent Program, Division of AIDS, NIAID, NIH: pNL4-3.Luc.R-E- from Dr. Nathaniel Landau, (34, 90)] and a VSV-G encoding plasmid using the calcium phosphate technique. Pseudovirus containing supernatants were harvested two days after transfection and cellular debris was removed by centrifugation at 300 × g for 5 min. HIV-1 p24 concentration was estimated by antigen enzyme linked immunosorbent assay (ELISA) performed by the Institute for Clinical and Molecular Virology. Pseudovirus containing supernatants were stored in 1 ml aliquots at -80°C until use.

3.2.4.4 Infection of Lymphocytes with HIV-1 and Stimulation with GlcN

Jurkat and T1 cells were cultured to a density of 1.5×10^6 cells/ml and split 1:2 the day prior to infection. Cells were infected with VSV-G *env* pseudotyped HIV-1$_{NL4-3LucR-E-}$ for 4 h at 37°C, using a concentration of 30 ng p24 per 1×10^6 cells. The cells were centrifuged at 300 × g for 5 min to remove excessive pseudovirus and resuspended in 2 ml VLE-RPMI 1640 containing 2 mM L-glutamine and 10% FCS supplemented with the respective concentration of GlcN in triplicates. CD4$^+$ T cells were stimulated 6 days with 10 U/ml IL-2 and 10 µg/ml PHA-P prior to infection. The lymphocytes were infected with reporter HIV-1 for 4 h at 37°C, using a concentration of 30 ng p24 per 2×10^6 cells. Cells were centrifuged at 250 × g for 10 min to remove excessive pseudovirus and resuspended in 1.5 ml MLPC medium supplemented with the respective concentration of GlcN in triplicates. After incubation for 24 h, cells were collected, washed once with PBS, lysed in 1× passive lysis buffer and assayed for luciferase activity as described below.

3.2.4.5 Cell Transfection

a) Transfection of HEK 293T Cells

HEK 293T cells were seeded 24 h prior to transfection in 1.5 ml medium at a density of 3×10^5 cells per well in a 6-well plate. Cells were transiently cotransfected *via* the calcium phosphate method in triplicates with 0.5 µg pXP1-LTR-Sp1$_{wt}$-Luc or 0.5 µg pEF1α-Luc, along with 0.75 µg pcDNA-Sp1 and/or 0.25 µg pCEP4-Tat and 1 µg pcDNA-OGT (unless otherwise indicated). The total DNA amount was adjusted to 2.5 µg with the control plasmid pcDNA4. For RNA interference experiments, 0.22 μg (amount corresponds to 10 nM) siRNA oligonucleotides specific for human OGT (wtOGT siRNA), human Sp1 (Sp1-siRNA), human GAPDH or a non-targeting control-siRNA (control-siRNA) were cotransfected. 24 h later, cells were washed once with PBS, lysed in 200 μl 1× passive lysis buffer and assayed for luciferase activity.

Stable transfection of HEK 293T cells with pcDNA4-LTRwt-d2EGFP and pcDNA4-LTRmutSp1-d2EGFP was performed *via* the calcium phosphate method. Stably transfected cells were selected with 600 μg/ml Zeocin. After 10 days, single colonies were transferred to new culture plates. Fluorescence intensity of clones was analyzed by flow cytometry. For the final experiment, three clones of each construct were selected, which expressed comparable d2EGFP fluorescence intensities.

b) Transfection of Jurkat Cells

Electroporation of Jurkat cells was performed with the Nucleofector II. Therefore, cells were resuspended in 100 μl Nucleofector Solution V at a final concentration of 2×10^7 cells/ml and mixed with 50 μg/ml DNA (pcDNA4-EF1α-d2EGFP or pcDNA4-OGT). Electroporation was carried out with 100 μl cell-DNA suspension using program X-001. Immediately after transfection, cells were transferred to prewarmed RPMI 1640 medium supplemented with 10% FCS and 4 mM L-glutamine. 48 h post transfection, stably transfected Jurkat cells were selected with 400 μg/ml Zeocin. After initial selection for 48 h, cells were singularized by limited dilution into 96-well plates. After 13 days, d2EGFP expressing Jurkat clones were pooled, while OGT expressing Jurkat clones were cultivated separately.

c) Transfection of $CD4^+$ Primary T Lymphocytes with mRNA

T lymphocytes were electroporated as described previously (198). Briefly, cells were washed once with pure RPMI 1640 and once with OptiMEM without phenol red (Invitrogen) (all at room temperature). The cells were resuspended in OptiMEM at a concentration of 1×10^8 cells/ml. RNA was transferred to a 4-mm cuvette (Molecular Bioproducts, San Diego, CA, USA) at a final concentration of 150 μg/ml. A volume of 100 μl of cell suspension was added and immediately pulsed in a Gene Pulser Xcell. Pulse conditions for $CD4^+$ T lymphocytes were square-wave pulse, 500 V and 5 ms. Immediately after electroporation, cells were transferred to prewarmed MLPC medium.

3.2.5 Protein Chemistry

3.2.5.1 Determination of Protein Concentration

The protein concentration of cell lysates was determined in a microplate reader at 750 nm using a detergent compatible (DC) protein assay kit according to the manufacturer's instructions. The colorimetric measurement is based on the Lowry method (140).

3.2.5.2 SDS-Polyacrylamide Gel Electrophoresis (SDS-PAGE)

Proteins were separated according to their molecular weight under denaturing conditions *via* sodium dodecyl sulphate (SDS)-PAGE by Laemmli (132). The cell lysates were diluted either with 2× Laemmli buffer or with 6× protein loading dye and boiled for 10 min at 99°C.

The electrophoretic separation of proteins was carried out in discontinuous SDS-polyacrylamide gels (composed of 2 units: a 5% stacking gel pH 6.8 and a 7.5% separation gel pH 8.8) or in 4-20% tris glycine gradient SDS polyacrylamide gels. Gels were run at 20 mA and 30 mA per gel, respectively, in 1× tris glycine SDS running buffer. To determine the molecular weight, a prestained molecular weight standard PageRuler™ protein ladder was used, containing marker proteins with definite molecular weights (170, 130, 95, 72/70, 55, 43, 34, 26, 17 und 11 kDa). The following table displays the composition of the 5% stacking and 7.5% separation gels.

Table 19. Composition of the stacking and separation gels.

Component	Stacking gel 5%	Saparation gel 10%
ddH$_2$O	3.4 ml	4.8 ml
30% Acrylamid Mix (29:1)	0.83 ml	2.5 ml
1,5 M Tris-HCl pH 8.8	-	2.5 ml
1 M Tris-HCl pH 6.8	0.63 ml	-
10% SDS	50 µl	100 µl
10% APS	50 µl	100 µl
TEMED	5 µl	4 µl
Total volume	5 ml	10 ml

3.2.5.3 Western Blot Analyses

Proteins separated by SDS-PAGE were transferred to a polyvinylidene fluoride (PVDF)-membrane. Therefore, the gels were stacked on the membrane between several layers of gel-blotting paper, taking care that no air bubbles are built. The transfer of the proteins from the gel to the PVDF-membrane was conducted under wet conditions at 250 mA for 2 h in 1× tris glycine buffer and 3% methanol. Small proteins were blotted onto Immobilon-P^{SQ} transfer membrane, optimized for

Western blotting of proteins smaller than 20 kDa. The transfer of the proteins to the Immobilon-P^{SQ} transfer membrane was conducted under wet conditions at 250 mA for 1 h in 1× tris glycine SDS buffer and 20% methanol.

The membranes were blocked with 5% non-fat dry milk in PBS/Tween for at least 2 h, washed with PBS/Tween, and incubated with the primary antibody in 2.5% non-fat dry milk in PBS/Tween at room temperature for 2 hours. After washing, the membranes were incubated 45 min at room temperature with the secondary antibody in 2.5% non-fat dry milk in PBS/Tween (224, 236). In order to prevent cross-reactions with milk proteins, Western blot staining with the O-GlcNAc-recognizing antibody was performed as follows: the membranes were blocked with 10% blocking solution and the antibody was diluted in 0.5% blocking solution. Subsequent to the final washing step, chemiluminescence was detected according to the manufacturer's protocol.

The following primary antibodies were used: polyclonal rabbit anti-human Sp1 Pep2, polyclonal goat anti-human lamin A/C N-18, polyclonal rabbit anti-human OGT TI-14, polyclonal rabbit anti-human actin, monoclonal mouse anti-human histone H1, monoclonal mouse anti-GFP, monoclonal mouse anti-eukaryote O-GlcNAc CTD110.6, and monoclonal mouse anti-human GAPDH. The monoclonal rat anti-Tat IgG2a antibody (clone 1C9; 1:150) was produced by immunization of LOU/C rats with His-tagged purified recombinant Tat protein (50 µg) according to a previously described procedure (148). The secondary, horseradish peroxidase (HRP) coupled antibodies goat anti-rat IgG, donkey anti-rabbit IgG, sheep anti-mouse IgG, and rabbit anti-goat IgG were diluted 1:5,000.

3.2.5.5 Luciferase Reporter Gene Assay

Cell lysis was performed in 1× passive lysis buffer for 15 min at RT. The cell lysates were centrifuged at 10,000 × g at 4°C for 3 min. Firefly luciferase activity of the supernatants was measured using the luciferase assay system in a Luminoskan Ascent instrument. Each luciferase reporter gene assay was performed in triplicate (except Fig. 3, in duplicate) and repeated at least 2 times.

3.2.5.6 Flow Cytometry Analyses

HeLa-Tat-III/LTR/d1EGFP cells were treated with the respective GlcN concentrations for 5 h in triplicates. HEK 293T-LTRwt-d2EGFP and HEK 293T-LTRmutSp1-d2EGFP cells were treated with 16 mM GlcN for 20 h in duplicates. The cells were harvested by trypsin/EDTA, washed twice with FACS buffer and resuspended in FACS buffer. 1×10^4 viable GFP positive cells were analyzed for each sample using a FACSCalibur flow cytometer with CellQuest Pro software. The relative fluorescence intensity is given as a ratio of the geo mean of untreated *versus* GlcN-treated cells. The geo mean of untreated cells was defined as 100%.

3.2.5.7 Cellular Analysis

a) CytoTox 96 Non-Radioactive Cytotoxicity Assay

The CytoTox 96 non-radioactive cytotoxicity assay is a colorimetric assay, which quantitatively measures LDH, a stable cytosolic enzyme that is released upon cell lysis. Released LDH in culture supernatants is measured with a 30-minute coupled enzymatic assay that results in the conversion of a tetrazolium salt into a red formazan product. The amount of color formed is proportional to the LDH acitivty. Visible wavelength absorbance data are collected using a standard 96-well plate reader. The CytoTox 96 non-radioactive cytotoxicity assay was performed according to the manufacturer's instructions. Fifty microliters of the HEK 293T supernatants were transferred into a 96-well in duplicate and incubated with 50 µl of reconstituted substrate mix for 30 min in the dark at RT. Subsequently, 50 µl stop solution were added to each well and the extinction of the formazan product was measured at 490 nm. The intensity of the red color correlates with the amount of dead cells in the supernatant and thus, is inverse proportional to the cell viability. The culture medium background caused by phenol red and serum was corrected after substracting the medium values from the cell supernatant values. Furthermore, a positive control (bovine heart LDH) dilution series was measured as experimental control.

b) CellTiter 96 Non-Radioactive Cell Proliferation Assay

The CellTiter 96 non-radioactive cell proliferation assay is a modification of the MTT assay method described by Mosmann (163) and incorporates several improvements. The CellTiter 96 assay was performed by adding 12 µl of a premixed, optimized Dye Solution to culture wells of a 96-well plate, containing 100 µl cell suspension with 1×10^5 HIV-1 infected Jurkat or T1 cells or 2×10^5 HIV-1 infected $CD4^+$ primary T lymphocytes stimulated for 24 h with GlcN. During an incubation of 1-4 h, living cells convert the MTT tetrazolium component of the Dye Solution into a formazan product. Fifty microliters of the Solubilization/Stop Solution are then added to the culture wells to solubilize the formazan product, and the absorbance at 570nm is recorded using a 96-well plate reader. The amount of color formed correlates directly with the cell number and is thus proportional to the cell viability. The culture medium background caused by phenol red and serum was corrected after substracting the medium values from the cell suspension values.

3.2.5.9 Nuclear/cytosol Fractionation

Fractionation was performed with a nuclear/cytosol fractionation kit according to the manufacturer's protocol (196). For gel shift experiments, cells were counted, and the nuclear fractions were lysed in 10 µl nuclear extraction buffer per 1×10^6 HEK 293T cells in order to concentrate the fractions. The protein concentration was determined and equal protein amounts of each fraction were loaded onto the gel and analyzed by Western blot or applied to gel shift experiments.

3.2.5.10 Immunoprecipitation

2×10^6 HEK 293T cells were seeded in 9 ml medium in 10 cm cell culture dishes. The cells were transfected 24 h later (*via* the calcium phosphate method) with pcDNA4, pcDNA-Sp1, or pcDNA-Sp1 and pcDNA-OGT. The total transfected DNA amount was adjusted to 30 µg with the control plasmid pcDNA4. 24 h after transfection, cells were washed with PBS and subsequently lysed in IP-lysis buffer, supplemented with 1 tablet Complete Mini, EDTA-free protease inhibitor cocktail per 10 ml. The protein concentration was determined and 1 mg total protein in a maximal volume of 1 ml was used for the immunoprecipitation. The lysates were precleared by incubation with 100 µl Sepharose CL-6B at 4°C for 1 h. The Sp1 protein complex was immunoprecipitated at 4°C for 3 h with 30 µg anti-Sp1 Pep2 antibody covalently coupled to agarose beads. Beads coupled protein complexes were washed twice with 1 ml IP-lysis buffer and three times with IP-wash buffer. The immunoprecipitates were analyzed by SDS-PAGE and Western blot.

3.2.5.11 Electrophoretic Mobility Shift Assay (EMSA)

a) Hybridization of Oligos

The oligonucleotides were adjusted to a concentration of 100 pmol/µl. The hybridization was carried out in a volume of 20 µl with the following components:

        6 µl sense oligonucleotide
        6 µl antisense oligonucleotide
        1.6 µl 5 M NaCl
        6.4 µl ddH$_2$O

The hybridization was performed in a Thermocycler. First, oligonucleotides were denatured at 95°C for 10 min. Subsequently, slow hybridization was enabled due to a decrease in the temperature by 5°C every 15 min. This step was repeated till the temperature reached 40°C. Finally, hybridized oligonucleotides were diluted to a concentration of 3 pmol/µl.

b) Labeling of Oligos

Labeling of oligos was carried out in a radionuclide laboratory with radioactive labeled ATP (γ-^{32}P-ATP) for 1 h at 37°C:

> 2 µl double-stranded oligonucleotide (3pmol/µl)
> 3 µl 10 PNK buffer
> 4 µl γ-^{32}P-ATP (222 TBq/mMol)
> 2 µl T4-PNK
> 19 µl ddH$_2$O

To remove excessive nucleotides, radioactive labeled oligonucleotides were purified *via* NucTrap probe purification columns. Therefore, columns were first rehydrated with 70 µl 1× STE. Subsequently, the probe diluted with 40 µl ddH$_2$O was pipetted directly into the middle of the column. The solution was pushed through the column with the aid of a syringe and eluted with 70 µl 1× STE. Labelling efficiency was determined by measuring the probes in the scintillator. The concentration of the radioactive labelled probes was adjusted to 2×10^4 cpmA/µl with 1× STE.

c) Band Shift Assay

HEK 293T cells were transfected with the expression constructs for Sp1, OGT or both and nuclear proteins were isolated as already described. The electrophoretic mobility shift assay was essentially performed as described previously (165, 169). Briefly, binding reaction were incubated 45 min at 4°C and contained (final volume 15 µl):

> 10 mM Tris-HCl (pH 7.5)
> 80 mM NaCl
> 1 mM EDTA
> 1 mM DTT
> 5% glycerol
> 3 µg of poly(dI-dC)·poly(dI-dC)
> 10 µg of nuclear extract
> 4×10^4 cpmA ^{32}P-end-labeled double-stranded oligonucleotide probe.

The oligonucleotides corresponded either to the wild type HIV-1 LTR region containing the three binding sites for Sp1 (wt LTR-Sp1) or to the Sp1-binding sites mutated HIV-1 LTR (mut LTR-Sp1). After incubation for 45 min on ice, the protein-DNA complexes were resolved on non-denaturing 5% polyacrylamide gels run in 1× TBE buffer. The 5% polyacrylamide gel was composed as follows (total volume 60 ml):

> 10 ml 40% acrylamide (29:1)
> 3 ml 10× TBE
> 47 ml ddH$_2$O
> 500 µl 10% APS
> 50 µl TEMED

For competition experiments, a 10-, 20- or 50-fold molar excess of unlabeled competitor oligomers was added to the gel shift mixtures prior to the addition of the ^{32}P-labeled oligonucleotide probe and incubated for 30 min at 4°C.

For supershift assays, 1 µg, 2 µg or 5 µg antibodies directed against Sp1, O-GlcNAc or OGT were added to the gel shift mixture and incubated for 30 min at 4°C prior to the addition of the labeled probe. Supershift antibodies were Sp1 Pep2 sc-59X, O-GlcNAc CTD110.6, and OGT TI-14, as well as IgG [NF-κB p50 sc114X (Santa Cruz Biotechnology)] and IgM [Heparan sulfate 370255 (seikagaku corporation)] isotype controls.

3.2.6 Statistical Analyses

Statistical significances were calculated with the Student's t-test for paired samples using the SPSS 15.0 and 16.0 software for Microsoft Windows (SPSS Inc., Chicago, Illinois, USA). P-values smaller than 0.05 were considered statistically significant (*), P-values smaller than 0.01 highly significant (**), and P-values smaller than 0.001 highest significant (***).

4. Results

4.1 GlcN Inhibits HIV-1 Transcription in Lymphocytes

In order to evaluate whether *O*-GlcNAcylation affects HIV-1 replication, Jurkat and T1 were infected with HIV-1$_{NL4-3LucR-E-}$, a replication deficient HIV-1 clone pseudotyped with the VSV-G envelope protein. Upon integration, this recombinant virus expresses the firefly luciferase gene driven by the HIV-1 LTR promoter. Consequently, the luciferase activity in infected cells correlates with the rate of HIV-1 gene transcription.

Reporter-HIV-1-infected cells were either left untreated (0 mM) or treated for 24 h with increasing concentrations of GlcN (0.25 mM, 1 mM, 4 mM, and 16 mM) and luciferase activity was measured. GlcN significantly inhibited the HIV-1 gene transcription by more than 60% in Jurkat cells and 50% in T1 cells (Fig. 14A and B, solid line) without significantly decreasing cell viability (Fig. 14A and B, dashed line). Western blot analyses using an antibody recognizing *O*-GlcNAcylated proteins demonstrated increased *O*-GlcNAcylation upon GlcN treatment, whereas expression of *O*-GlcNAcylated proteins like OGT and non-*O*-GlcNAcylated proteins such as actin remained unchanged (Fig. 14C). Detection of GAPDH demonstrated that equal amounts of protein were loaded.

Similar results were obtained in primary CD4$^+$ T lymphocytes from two different donors (Fig. 14D and E), although HIV-1 replication in donor 2 (Fig. 14E) was inhibited to a lesser extent as compared to donor 1 (Fig. 14D). This difference is also reflected in lower amounts of *O*-GlcNAcylated proteins in the lysates of donor 2 as compared to donor 1 (compare Fig. 14F, upper panel, donor 1 and donor 2).

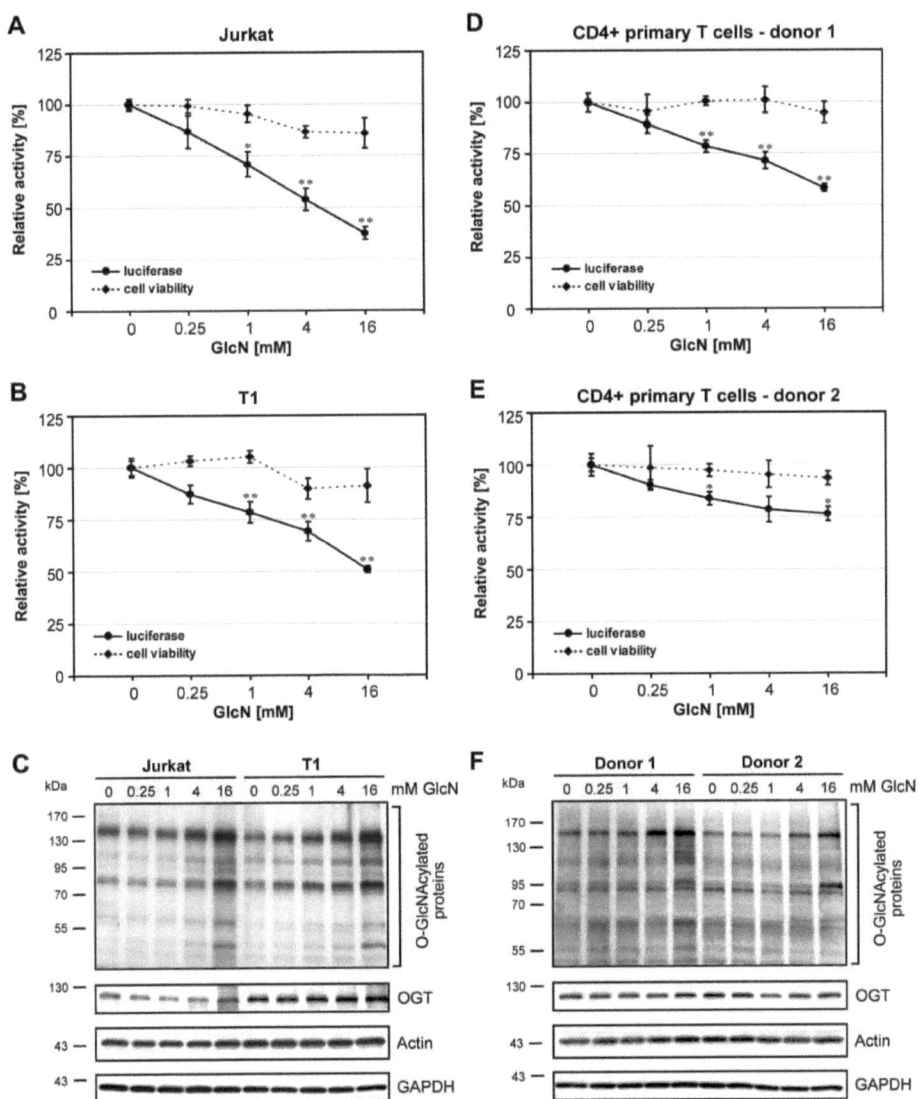

Fig. 14. GlcN inhibits HIV-1 transcription in T-lymphocytic cell lines and in primary CD4+ T cells. (A) Jurkat, (B) T1 cells and (D, E) primary CD4+ T cells from two different donors were infected with VSV-G *env* pseudotyped HIV-$1_{NL4-3LucR-E-}$ and cultured in the absence (0 mM) or presence of different concentrations of GlcN (0.25 mM, 1 mM, 4 mM, and 16 mM) for 24 h. Subsequently, HIV-1 LTR-driven luciferase activity was measured (solid line). The effect of GlcN on cytotoxicity and proliferation was monitored by MTT assay (dashed line). The results are presented in terms of percent activity of untreated control cells. The means ± standard deviations from triplicate determinations are indicated. P-values are calculated in comparison with control (0 mM GlcN): *, $P \leq 0.05$; **, $P \leq 0.01$. Western blot analyses of *O*-GlcNAcylated proteins in (C) Jurkat and T1 cells and in (F) primary CD4+ T cells. OGT served as a control for *O*-GlcNAcylated proteins, and actin for non-*O*-GlcNAcylated proteins. Staining of GAPDH demonstrated equal loading of proteins.

4.2 GlcN Represses HIV-1 LTR Promoter Activity in HeLa Cells

To exclude the possibility that reduced HIV-1 gene transcription is caused by impaired nuclear transport of the preintegration complex or reduced proviral integration, the impact of GlcN treatment on the HIV-1 LTR promoter activity was evaluated. Therefore, HeLa-Tat-III/LTR/d1EGFP cells stably expressing d1EGFP under the control of the HIV-1 LTR were analyzed by flow cytometry. The cells were either left untreated (0 mM) or treated with increasing concentrations of GlcN (0.25 mM, 1 mM, 4 mM, and 16 mM) for 5 h. Subsequently, fluorescence intensity was measured. GlcN treatment inhibited the HIV-1 LTR-triggered d1EGFP expression in a dose-dependent manner, as detected by the shift of the fluorescence emission peak to lower fluorescence values (Fig. 15A). Quantification of the different values demonstrated that GlcN significantly decreased the HIV-1 LTR activity by more than 60% (Fig. 15B). Comparable results were obtained after treatment of HeLa-Tat-III/LTR/d1EGFP cells with the UDP-GlcNAc analog streptozotocin (data not shown), which inhibits the O-GlcNAcase activity and thereby increases O-GlcNAcylation (192).

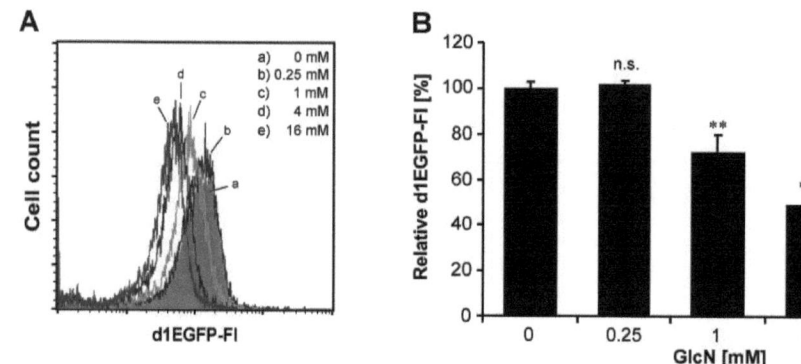

Fig. 15. GlcN inhibits HIV-1 LTR promoter activity in HeLa cells. HeLa-Tat-III/LTR/d1EGFP cells were either left untreated [0 mM (a)] or stimulated for 5 h with increasing concentrations of GlcN [0.25 mM (b), 1 mM (c), 4 mM (d), and 16 mM (e)]. The HIV-1 LTR activity was assessed by measuring the d1EGFP fluorescence intensity (FI) with flow cytometry analyses. **(A)** The shift of the fluorescence emission peak and **(B)** the respective bar diagram of the different geo mean values compared to unstimulated control cells are shown. The means ± standard deviations calculated from triplicate determinations are indicated. P-values are given for comparison with control: n.s., not significant; **, $P \leq 0.01$; ***, $P \leq 0.001$.

To demonstrate O-GlcNAcylation of cellular proteins in the presence of GlcN, cytosolic and nuclear fractions of control-treated (0 mM) *versus* GlcN-treated (16 mM) HeLa-Tat-III/LTR/d1EGFP cells were analyzed by Western blot (Fig. 16). In both fractions, several proteins showed increased O-GlcNAcylation upon GlcN treatment (Fig. 16, upper panels, arrowheads), whereas the expression of OGT (example of an O-GlcNAcylated protein) and of histone H1 (non-O-GlcNAcylated protein) remained unchanged (Fig. 16, middle panels). Successful fractionation of cytoplasmic and nuclear proteins and loading of equal protein amounts were verified by staining of lamin A/C and GAPDH (Fig. 16, lower panels). Altogether, these results demonstrate that GlcN treatment enhances O-GlcNAcylation of cytoplasmic and nuclear proteins and that this correlates with decreased gene expression from the HIV-1 LTR promoter.

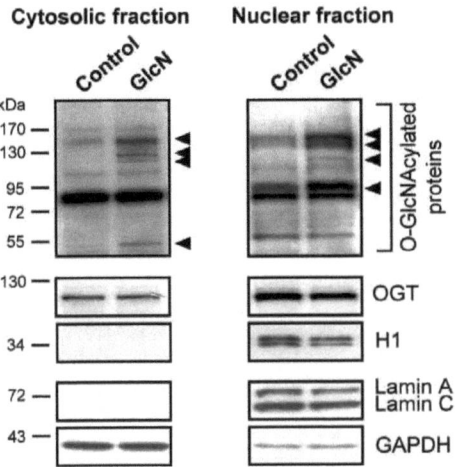

Fig. 16. GlcN increases *O*-GlcNAcylation of cytoplasmic and nuclear proteins in HeLa cells. Cytosolic and nuclear fractions of unstimulated (control) or stimulated (16 mM GlcN) HeLa-Tat-III/LTR/d1EGFP cells were analyzed by Western blot to detect *O*-GlcNAcylated proteins. Arrowheads mark proteins whose *O*-GlcNAcylation patterns are increased upon GlcN treatment. Staining of OGT (example of an *O*-GlcNAcylated protein) and histone H1 (non-*O*-GlcNAcylated protein) showed that protein expression levels were not altered. Staining of GAPDH and lamin A/C demonstrated successful fractionation.

4.3 OGT Inhibits HIV-1 Transcription in Primary Lymphocytes

Since OGT is the only known enzyme to mediate *O*-GlcNAcylation, the effect of OGT on HIV-1 transcription in primary lymphocytes was investigated. To this goal, primary $CD4^+$ T lymphocytes from two different donors were infected with pseudotyped reporter HIV-$1_{NL4-3LucR-E-}$. 36 h post infection the cells were transfected with *in vitro* transcribed polyadenylated mRNA encoding either EGFP or OGT. OGT expression levels were maximal at 8 h after transfection (data not shown) and therefore the cells were harvested at this time point. Reporter analyses showed that overexpression of OGT inhibited HIV-1 transcription (Fig. 17A). OGT-mediated inhibition of the HIV-1 LTR was to a similar extent as inhibition mediated by treatment with 16 mM GlcN (compare Fig. 17A with Fig. 14D and E). The inhibition was slightly stronger in the cells of donor 2 as compared to donor 1, which correlated well with the increased OGT expression and *O*-GlcNAcylation level observed in the cells of donor 2 (Fig. 17B, upper panels, compare donor 1 and donor 2).

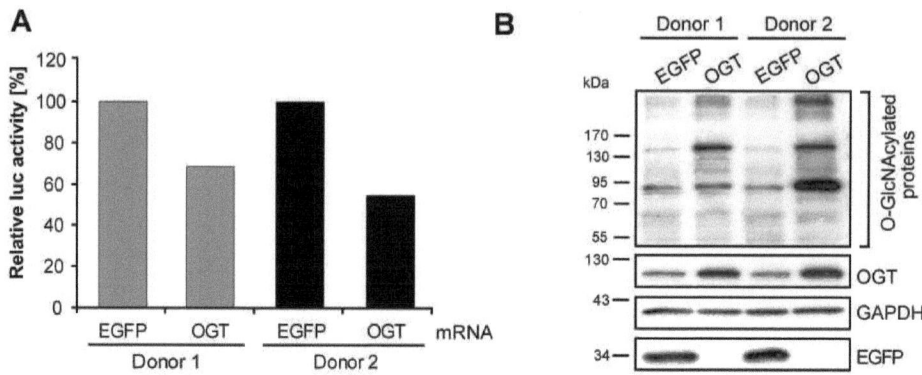

Fig. 17. OGT inhibits HIV-1 LTR promoter activity in infected primary CD4⁺ T cells. Primary CD4⁺ T cells from two different donors were infected with VSV-G *env* pseudotyped HIV-1$_{NL4-3LucR-E-}$. 36 h post infection, cells were electroporated with *in vitro* transcribed polyadenylated mRNA encoding either EGFP or OGT. **(A)** 8 h post electroporation, luciferase activity was measured. **(B)** *O*-GlcNAcylation pattern and overexpression of EGFP and OGT were verified by Western blot. Staining of GAPDH demonstrated equal loading of proteins.

4.4 The Sp1-Binding Sites Are Necessary and Sufficient for the Inhibition of HIV-1 LTR Activity by OGT

Assuming that *O*-GlcNAcylation alters the activity of transcription factors, Sp1 and NF-κB were prime candidates for the *O*-GlcNAc-mediated inhibition of the HIV-1 LTR. Binding sites for both factors constitute the core promoter/enhancer of HIV-1, and both have been described to be modified by *O*-GlcNAc (69, 101). In order to prove this hypothesis, luciferase reporter assays were performed using LTR reporters lacking the upstream negative regulatory region, but containing the NF-κB- and/or Sp1-binding sites (either combined or isolated), as well as a TATA-box and the transactivation response (TAR) element for Tat-mediated activation (Fig. 18A, LTR-κB-Sp1wt, LTR-κB-Sp1mut, LTR-Sp1wt). HEK 293T cells were cotransfected with the reporter constructs and a Tat encoding plasmid, along with an OGT encoding plasmid or the control plasmid pcDNA4 (Fig. 18B). Overexpression of OGT inhibited the Tat-induced activity of the promoters containing functional Sp1-binding sites (Fig. 18B, LTR-κB-Sp1$_{wt}$ and LTR-Sp1$_{wt}$). In contrast, the activity of the promoter containing the NF-κB motif but mutated Sp1-binding sites was increased upon overexpression of OGT (Fig. 18B, LTR-κB-Sp1$_{mut}$). The transient expression of Tat itself was not significantly affected by overexpression of OGT (Fig. 18B, lower panels). These findings suggest that the presence of functional Sp1-binding sites is crucial for the inhibitory effect of OGT on the HIV-1 LTR.

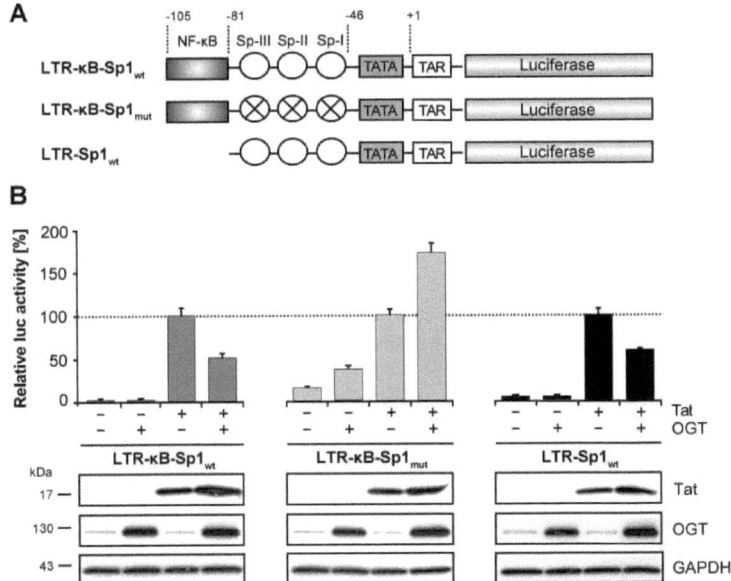

Fig. 18. The presence of Sp1-binding sites in the HIV-1 LTR promoter is required for OGT-mediated inhibition. (**A**) Schematic representation of the promoter constructs used in the luciferase assay in (B). The truncated promoters lack the upstream negative regulatory region, but contain the NF-κB and/or Sp1-binding sites (either combined or isolated), as well as a TATA-box and the transactivation response (TAR) element for Tat-mediated activation. Crossed circles represent mutated Sp1-binding sites. Numbers reflect the positions in the wild type LTR_{LAI}. (**B**) Promoter activities were detected by measuring the luciferase activity in the cell lysates of HEK 293T cells transiently transfected with the reporter constructs (LTR-κB-Sp1$_{wt}$, LTR-κB-Sp1$_{mut}$, LTR-Sp1$_{wt}$) together with plasmids coding for OGT and Tat as indicated. The total DNA amount was adjusted with pcDNA4. The values were normalized to the total amount of protein and are presented in terms of percent of the corresponding Tat-induced promoter activity. The means ± standard deviations from triplicate determinations are shown. Western blot analyses of Tat and OGT revealed the expected expression pattern. GAPDH staining demonstrated that equal amounts of protein were loaded.

In a next step, the role of the Sp1-binding sites in the full-length HIV-1 LTR integrated into the cellular genome was analyzed. To this goal HEK 293T cells were stably transfected with plasmids expressing d2EGFP either under the control of the full-length wild type HIV-1 LTR (Fig. 19A, LTRwt) or under the control of the full-length LTR containing mutated Sp1-binding sites (Fig. 19A, LTRmutSp1). Three independent HEK 293T cell clones with similar fluorescence intensity were either left untreated or were stimulated with 16 mM GlcN. Subsequently, the promoter activities were measured by flow cytometry. The inactivation of the Sp1-binding sites strongly decreased HIV-1 LTR activity (Fig. 19B, compare fluorescence intensities in the bar diagram). However, GlcN treatment significantly decreased the fluorescence intensity only in cells expressing d2EGFP under the control of the wild type HIV-1 LTR, but had no significant effect on cells expressing d2EGFP under the control of the Sp1-mutated HIV-1 LTR (Fig. 19B, bar diagram). Western blot analyses demonstrated that GlcN treatment similarly increased the *O*-GlcNAcylation pattern in both cell types, although OGT expression was not altered (Fig. 19B, right panels). Altogether these results demonstrate that the Sp1-binding sites are required for the *O*-GlcNAc-mediated inhibition of the full length HIV-1 promoter integrated into the cellular genome.

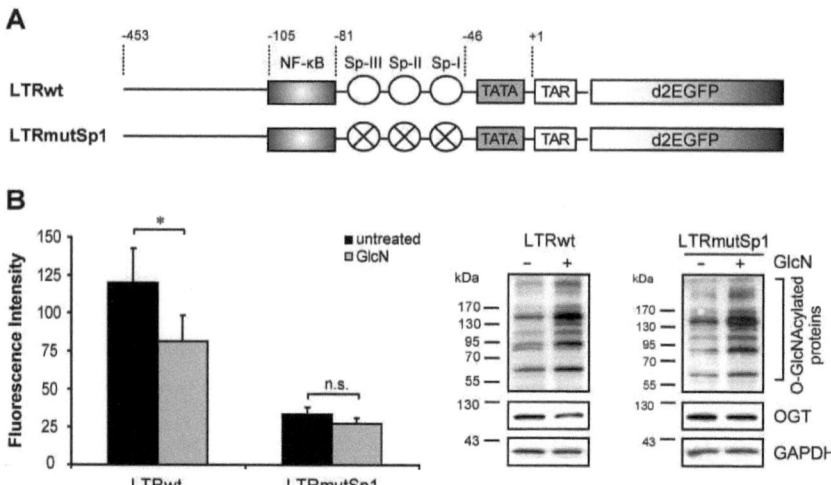

Fig. 19. The presence of Sp1-binding sites in the integrated HIV-1 LTR promoter is essential for *O*-GlcNAc-mediated inhibition. (A) Schematic representation of the full-length wild type (LTRwt) and full-length Sp1-mutated promoter (LTRmutSp1) constructs used for the stable transfection of HEK 293T cells. (B) HEK 293T cells stably expressing d2EGFP either under the control of the LTRwt or LTRmutSp1 were analyzed upon GlcN treatment (16 mM) by flow cytometry. The mean fluorescence intensity is represented in the bar diagram. The means ± standard deviations were calculated from duplicate determinations of three independent clones with similar fluorescence intensity. *P*-values are given for comparison with control: *, $P \leq 0.05$; n.s., not significant. Western blot analyses of one representative clone verified the increased *O*-GlcNAcylation pattern and unchanged expression of OGT upon GlcN treatment. GAPDH staining demonstrated that equal amounts of protein were loaded.

In an attempt to analyze the relevance of the individual Sp1-binding sequences in the *O*-GlcNAc-mediated inhibition of the HIV-1 LTR, a set of reporter plasmids containing the proximal region of HIV-1 LTR with either wild-type or mutated Sp1-binding sites (Fig. 20A) were employed in the cotransfection assays in HEK 293T cells (Fig. 20B). Mutation of the most promoter-distal Sp1-binding site Sp-III halved the inducibility of the HIV-1 LTR promoter by Sp1 (Fig. 20B, LTR-Sp1$_{mut-Sp-III}$), whereas inactivation of the promoter-medial and promoter-proximal Sp1-binding sites Sp-II or Sp-I markedly decreased the Sp1-mediated activation of the HIV-1 LTR promoter to 10% and 12.5%, respectively (Fig. 20, LTR-Sp1$_{mut-Sp-II}$ and LTR-Sp1$_{mut-Sp-I}$). Furthermore, mutation of both medial and proximal Sp1-binding sites completely abolished promoter activity (Fig. 20, LTR-Sp1$_{mut-Sp-I+II}$). This is in agreement with previous results (9), describing that the two proximal Sp1 motifs are more important than the promoter-distal Sp1-binding site for the transactivation of the HIV-1 LTR by Tat.

Accordingly, inactivation of the distal Sp1 motif did not affect the inhibition by OGT (Fig. 20B, LTR-Sp1$_{mut-Sp-III}$). Interestingly, the mutation of either medial or proximal Sp1-binding site almost completely abrogated the inhibition by OGT (Fig. 20, LTR-Sp1$_{mut-Sp-II}$ and LTR-Sp1$_{mut-Sp-I}$). Moreover, when both medial and proximal binding sites were mutated, a 3fold increase of promoter activity was observed in the presence of OGT (Fig. 20, LTR-Sp1$_{mut-Sp-I+II}$). These results indicate that the promoter-medial and -proximal Sp1 binding sites mediate the inhibitory effect of OGT on the HIV-1 LTR.

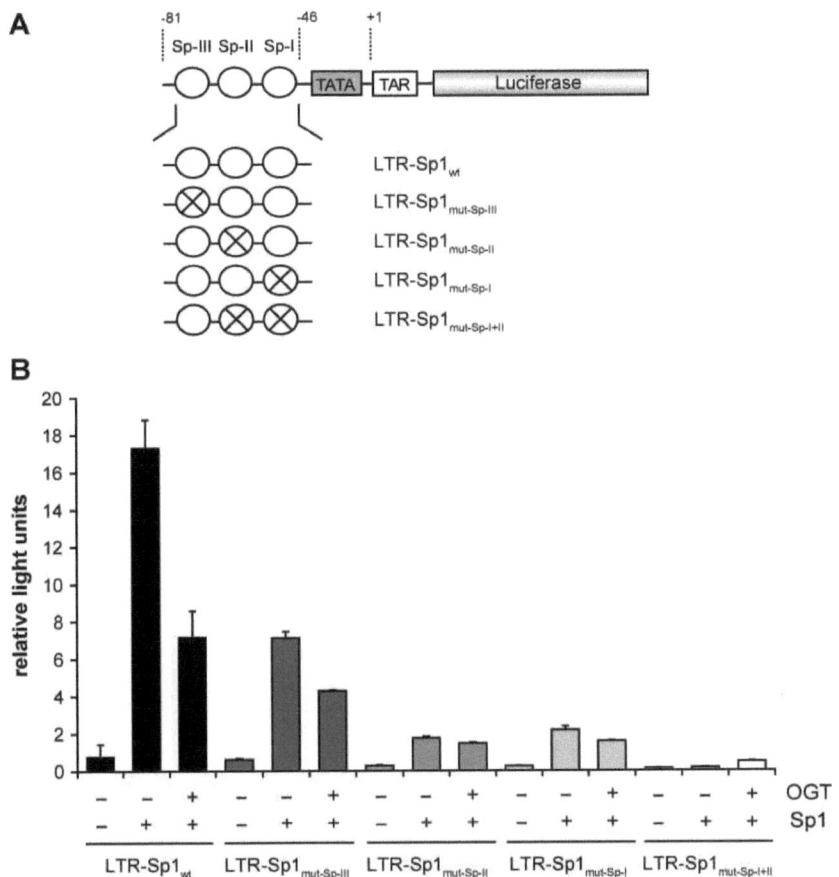

Fig. 20. The promoter-medial and promoter-proximal Sp1-binding sites mediate the inhibition of the HIV-1 LTR by OGT. (A) Schematic representation of the HIV-1 LTR constructs used in the luciferase assay in (B). The truncated promoters contain the proximal region of the HIV-1 LTR consisting of the Sp1-binding sites, the TATA-box and the TAR element. Crossed circles represent mutated Sp1-binding sites. Numbers reflect the positions in the wild type LTR$_{LAI}$. **(B)** HEK 293T cells were cotransfected with the reporter constructs (LTR-Sp1$_{wt}$, LTR-Sp1$_{mut-Sp-III}$, LTR-Sp1$_{mut-Sp-II}$, LTR-Sp1$_{mut-Sp-I}$ and LTR-Sp1$_{mut-Sp-I+II}$) together with expression plasmids encoding Sp1 and OGT as indicated. The total DNA amount was adjusted with pcDNA4. The values were normalized to the total amount of protein and are presented in terms of relative light units. The means ± standard deviations from triplicate determinations are shown.

4.5 Sp1 Is Necessary for the Inhibitory Effect of OGT on the HIV-1 LTR

Since the presence of the Sp1-binding sites is crucial for the inhibitory effect of OGT on the HIV-1 LTR promoter, it was investigated whether this is also the case for the Sp1 protein itself. To this goal, the sensitivity of the Tat-induced LTR-Sp1$_{wt}$ promoter to OGT was analyzed in HEK 293T cells in the absence and presence of Sp1 protein. Sp1 was recombinantly overexpressed and its expression was depleted by a specific short interfering RNA (Sp1-siRNA). A non-targeting siRNA (control-siRNA) was used as control. Cotransfection of the control siRNA did not affect OGT-mediated inhibition of the LTR-Sp1wt (Fig. 21, compare bars 3 and 4). Tat-mediated activation of the LTR-Sp1$_{wt}$ was less pronounced after knockdown of Sp1 (Fig. 21, compare relative light unit values given above the bars 3 and 7). However, OGT clearly did not inhibit LTR-Sp1$_{wt}$ promoter activity under these conditions (Fig. 21, compare bars 7 and 8). Thus, in addition to the Sp1-binding sites, the presence of the Sp1 protein itself is also necessary for the OGT-mediated inhibition of the HIV-1 LTR activity.

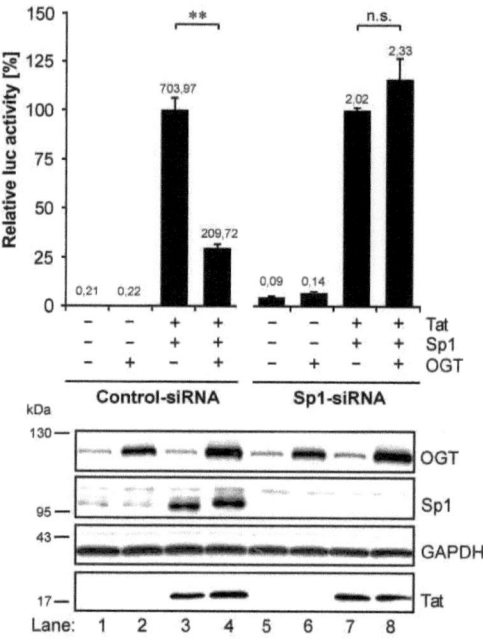

Fig. 21. Sp1 is crucial for the inhibitory effect of OGT on the HIV-1 LTR. HIV-1 LTR activity was measured by luciferase assay after cotransfection of HEK 293T cells with the LTR-Sp1$_{wt}$ reporter construct along with Tat and Sp1 encoding plasmids in the absence or presence of an OGT encoding vector together with 10 nM control siRNA or Sp1-targeting siRNA. Total DNA amount was adjusted with pcDNA4. The values above the bars indicate the means of the relative light units normalized to the total protein content ± standard deviations from triplicate determinations. The results were normalized for each siRNA dataset individually and are presented in terms of percent of control. P-values are given for comparison with control: **, $P \leq 0.01$; n.s., not significant. Western blot analyses of the lysates verified the knockdown of Sp1 by siRNA and demonstrated that the expression of OGT and Tat remains unaffected upon silencing of Sp1. The GAPDH staining demonstrated equal loading of proteins.

4.6 *O*-GlcNAcylation of Sp1 Selectively and Dose-Dependently Inhibits the HIV-1 LTR

To investigate the role of Sp1 in the OGT-mediated inhibition of the HIV-1 LTR in more detail, *O*-GlcNAcylation of Sp1 was explored in conjunction with OGT overexpression (Fig. 22). For this purpose, Sp1 was expressed alone or in combination with OGT (Fig. 22, Input). Immunoprecipitation experiments demonstrated that *O*-GlcNAcylation of Sp1 was substantially increased when OGT was overexpressed (Fig. 22, Sp1-IP). Coimmunoprecipitations revealed that OGT physically interacts with Sp1 under these conditions (data not shown).

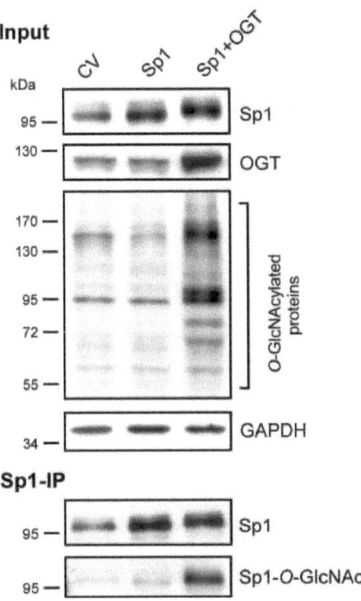

Fig. 22. Sp1 is strongly *O*-GlcNAcylated upon overexpression of OGT. HEK 293T cells were transfected with control vector (CV), Sp1, or Sp1 and OGT expressing vectors. Total DNA amount was adjusted with pcDNA4. Expression of transfected plasmids and *O*-GlcNAcylation were assessed *via* Western blot of cell lysates (Input). Immunoprecipitation of Sp1 (Sp1-IP) and its *O*-GlcNAcylation were verified by Western blot of Sp1-immunoprecipitates with anti-Sp1 and anti-*O*-GlcNAc antibodies, respectively.

In order to determine whether *O*-GlcNAcylation of Sp1 inhibits solely HIV-1 LTR activity or transcription in general, luciferase reporter assays were performed using the LTR-Sp1$_{wt}$ and an EF1α promoter construct. The latter triggers the expression of the housekeeping gene EF1α and harbors Sp1-binding sites. Unlike the LTR-Sp1$_{wt}$ promoter, the EF1α promoter has no TAR-element. In order to ensure a comparable induction, both reporter promoters were activated with Sp1 alone (Fig. 23A). In agreement with the results above, overexpression of OGT significantly inhibited Sp1-induced expression of the LTR-Sp1$_{wt}$ (Fig. 23A, LTR-Sp1$_{wt}$). In contrast, the activity of the EF1α promoter was not repressed, but significantly increased by OGT (Fig. 23A, EF1α), suggesting that OGT inhibits selectively the HIV-1 LTR promoter and does not generally repress Sp1-regulated gene expression.

Furthermore, a luciferase reporter assay was used to evaluate whether the inhibitory effect of OGT on the HIV-1 LTR promoter is dose-dependent. To this end, the reporter plasmid LTR-Sp1$_{wt}$ was cotransfected with Sp1 and Tat encoding constructs in HEK 293T cells together with increasing amounts of an OGT encoding plasmid (0.05 µg, 0.2 µg, 1 µg). OGT clearly inhibited dose dependently the HIV-1 LTR promoter activity (Fig. 23B).

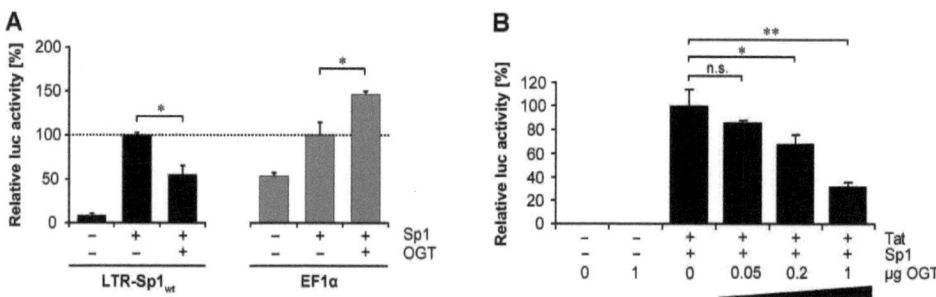

Fig. 23. *O*-GlcNAcylation of Sp1 selectively inhibits HIV-1 LTR promoter activity in a dose-dependent manner. (A) The effects of OGT on the LTR-Sp1$_{wt}$ and the control promoter EF1α in HEK 293T cells cotransfected with the reporter and an Sp1 encoding plasmid were analyzed by luciferase assay. (B) Luciferase activity of HEK 293T cells transiently cotransfected with the LTR-Sp1$_{wt}$ reporter construct along with Tat and Sp1 encoding plasmids and increasing concentrations of an OGT encoding vector (0.05 µg, 0.2 µg, 1 µg). The total DNA amount in (A) and (B) was adjusted with pcDNA4. The values were normalized to the total amount of protein and are presented in terms of percent of control (A, Sp1-induced; B, Tat and Sp1-induced promoter activity). The means ± standard deviations from triplicate determinations are indicated. *P*-values are given for comparison with control: n.s., not significant; *, $P \leq 0.05$; **, $P \leq 0.01$.

4.7 OGT Does Not Inhibit Expression and DNA-Binding Affinity of Sp1

The inhibitory effect of OGT on the HIV-1 LTR may rely on impaired nuclear translocation or decreased DNA-binding affinity of Sp1 upon *O*-GlcNAcylation. Therefore, the impact of OGT on the nuclear translocation of Sp1 was investigated by comparing the nuclear amount of Sp1 in cells expressing endogenous or increased levels of OGT. The amount of nuclear Sp1 was not decreased upon overexpression of OGT (Fig. 24), but rather increased, suggesting that the observed decrease in the HIV-1 LTR promoter activity was not due to impaired nuclear translocation of Sp1.

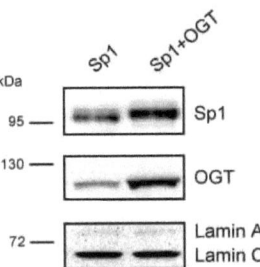

Fig. 24. OGT does not interfere with Sp1 expression. HEK 293T cells were transfected with an Sp1 encoding plasmid, along with an OGT encoding plasmid or control plasmid. Protein expression was detected in nuclear protein extracts via Western blot. Staining of lamin A/C was used as a loading control.

To investigate whether OGT interferes with the DNA-binding affinity of Sp1, gel shift assays were carried out with templates corresponding to the Sp1-binding sites in the HIV-1 LTR promoter (Fig. 25A, wt LTR-Sp1). No decrease in the formation of protein-oligonucleotide complexes was detected upon overexpression of OGT (Fig. 25A, compare lane 3 and 5). As a specificity control, gel shift experiments were performed using templates containing mutated Sp1-binding sites (Fig. 25A, mut LTR-Sp1). The specific Sp1 complex (Sp1-C) was not detectable with these oligonucleotides (Fig. 25A, lanes 4 and 6).

The specificity of the Sp1-C was confirmed by competition experiments with increasing molar excess of unlabeled wt LTR-Sp1 (Fig. 25B, lanes 8 to 10) and mut LTR-Sp1 oligonucleotides (Fig. 25B, lanes 11 to 13). In addition, the presence of O-GlcNAc modified Sp1 in the Sp1-C was confirmed by supershift analyses using increasing amounts of anti-Sp1 (Fig. 25C, lanes 15 to 17) and anti-O-GlcNAc antibodies (Fig. 25C, lanes 18 to 20). Both antibodies shifted the Sp1-C almost completely (Fig. 25C, lanes 17 and 20), suggesting that most of the DNA-bound Sp1 is O-GlcNAcylated in OGT overexpressing cells. Addition of an anti-OGT antibody did not shift the Sp1-C (Fig. 25C, lanes 21 to 23), indicating that OGT does not interact with Sp1 at the promoter site. Isotype control anti-IgG (Fig. 25C, lane 24) and anti-IgM antibodies (Fig. 25C, lane 25) had no effect on the Sp1-C. These results demonstrate that increased O-GlcNAcylation of Sp1 does neither inhibit the ability of Sp1 to translocate into the nucleus, nor the DNA-binding affinity of Sp1.

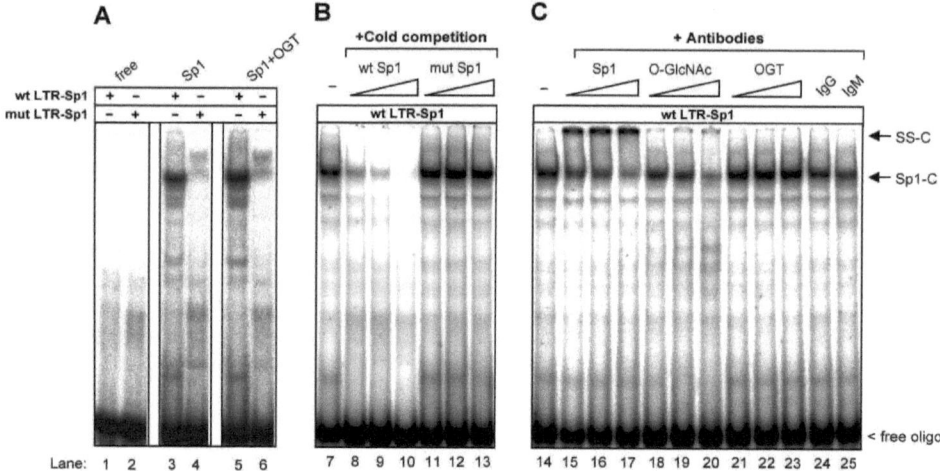

Fig. 25. OGT does not interfere with the DNA binding affinity of Sp1. (A) Electrophoretic mobility shift assays were carried out with ^{32}P-end-labeled double-stranded oligonucleotides corresponding to the Sp1-binding sites in the HIV-1 LTR promoter (wt LTR-Sp1) or with oligonucleotides containing mutated Sp1-binding sites in order to prevent binding (mut LTR-Sp1). One representative gel shift assay out of three is shown. Reactions were performed either without nuclear extracts (free), or with lysates from cells transfected with an Sp1 encoding plasmid alone (Sp1) or in combination with an OGT encoding plasmid (Sp1+OGT). The Sp1-oligonucleotide complex is indicated as Sp1-C. (B) Competition experiments were carried out with 10-, 20- and 50-fold molar excess of unlabeled wild type (wt LTR-Sp1) or mutated (mut LTR-Sp1) oligonucleotides. (C) Supershift analyses were performed using 1 μg, 2 μg or 5 μg anti-Sp1, anti-O-GlcNAc or anti-OGT antibodies as well as 5 μg anti-IgG or IgM as isotype control.

4.8 Sp1-O-GlcNAcylation Is Necessary for the Inhibition of the HIV-1 LTR

To confirm that the inhibition of the HIV-1 LTR is dependent on Sp1-O-GlcNAcylation, a specific siRNA was used to deplete the expression of OGT. Specificity of the siRNA was verified with an OGT rescue mutant (resOGT) encoding plasmid, which contains six silent nucleotide exchanges in the siRNA-binding site (Fig. 26A, nucleotide exchanges marked in bold). Western blot analyses proved that overexpression of wild type OGT (wtOGT) was inhibited in HEK 293T cells cotransfected with the wtOGT specific siRNA (Fig. 26B, upper panel, lanes 2 and 4), whereas expression of resOGT was not affected (Fig. 26B, upper panel, lanes 3 and 5). Staining of O-GlcNAc-modified proteins served as additional control for the functionality of resOGT (Fig. 26B, middle panel).

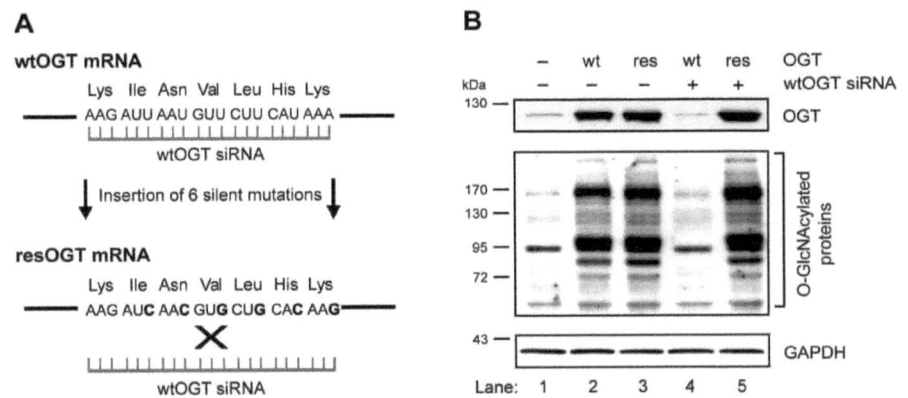

Fig. 26. Construction of an OGT rescue mutant and siRNA-mediated knockdown. (A) Generation of a plasmid encoding an OGT rescue mutant (resOGT) in order to escape silencing by wtOGT-targeting siRNA: 6 silent mutations (bold) were introduced into the siRNA-binding sequence of OGT. (B) Plasmids encoding wild type (wt) and rescue mutant (res) OGT were transfected together with or without wtOGT siRNA. OGT expression was detected by Western blot analysis. Staining of O-GlcNAcylated proteins was used as a control for the functionality of the rescue mutant. Immunodetection of GAPDH demonstrates that equal amounts of protein were loaded.

In order to investigate whether knockdown of OGT has a cytotoxic effect, HEK 293T cells were transfected with plasmids encoding either GFP, wild type OGT (wtOGT) or the rescue mutant OGT (resOGT) together with non-targeting control siRNA or wtOGT-targeting siRNA. Cell growth and viability were monitored at different time points after transfection (24, 48, 72 and 96 h) using phase contrast microscopy and a cytotoxicity assay measuring the LDH activity in the cell supernatants. In addition, protein expression of OGT was analyzed in Western blots (Fig. 27). These experiments confirmed that transfection efficiency was at about 70% (Fig. 27, GFP fluorescence) and OGT expression was strongly downregulated by siRNA (Fig. 27, Western blot analyses). Of note, no significant alterations in cell growth and viability were observed at all time points (Fig. 27, phase contrast pictures and LDH release), arguing that depletion of OGT does not alter cell viability.

Fig. 27. Knockdown of OGT does not alter cell viability. HEK 293T cells were transfected with plasmids encoding GFP, wtOGT or resOGT together with 10 nM non-targeting control or OGT-targeting siRNA. After 24 h, 48 h, 72 h and 96 h GFP expression, cell growth and viability were monitored *via* fluorescence and phase contrast microscopy and LDH assay (bar diagram), respectively. The LDH release was calculated relatively (%) of control (GFP) transfected cells and is given in means ± standard deviations from duplicate determinations. Western blot analyses of the cell lysates confirm overexpression and knockdown of OGT. Equal expression of GAPDH at all time points reflected equal input amounts.

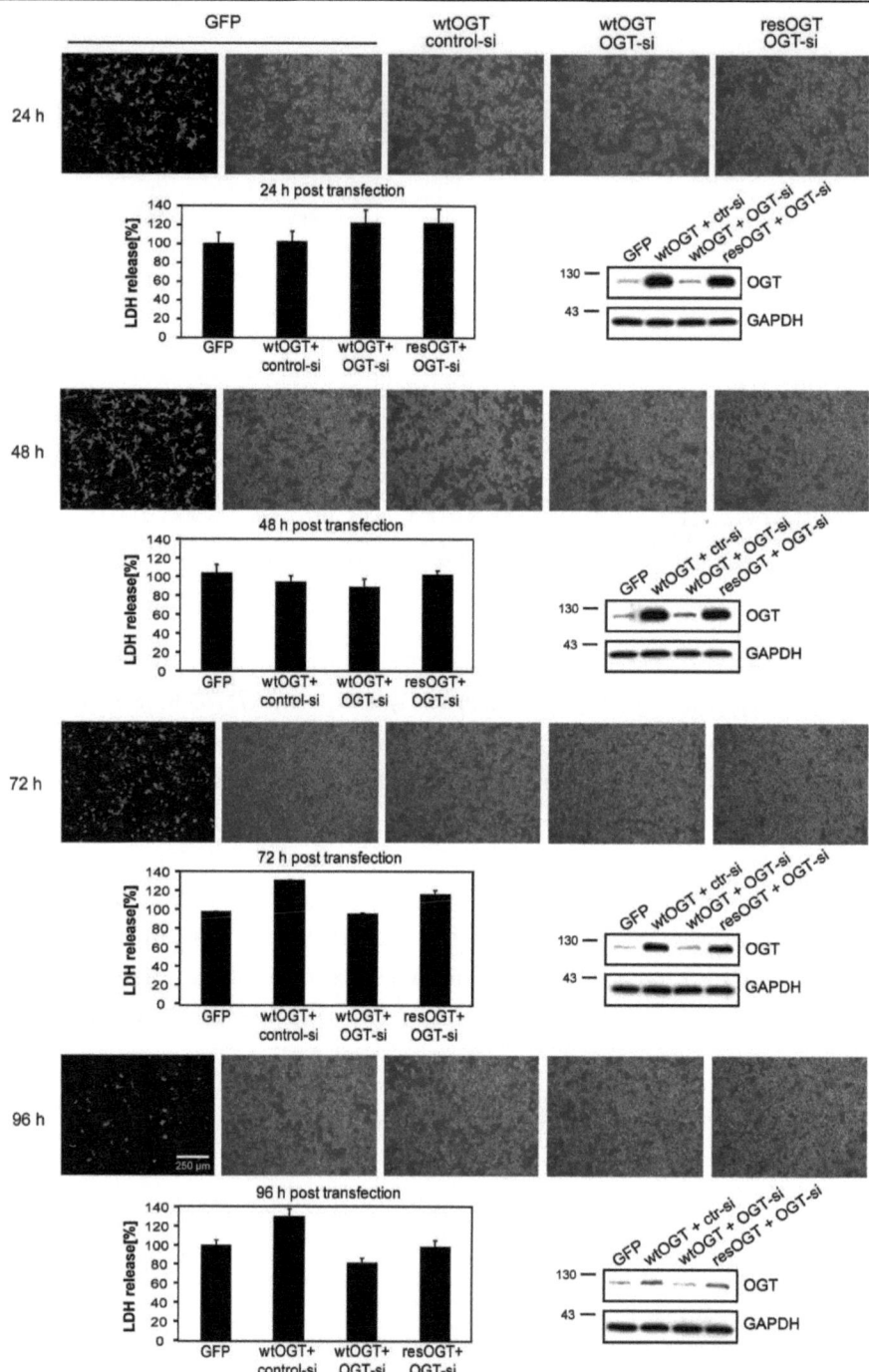

The established siRNA approach was applied to test the effect of Sp1-*O*-GlcNAcylation on the HIV-1 LTR. The promoter activity (Fig. 28, bar 4) was strongly reduced after overexpression of wtOGT or resOGT (Fig. 28, bars 5 and 6). Cotransfection of a control siRNA targeting GAPDH had no impact on the HIV-1 LTR activity (Fig. 28, bars 7 and 8). In contrast, knockdown of wtOGT restored the HIV-1 LTR activity (Fig. 28, bar 9), whereas resOGT escaped silencing and was still able to suppress the promoter activity (Fig. 28, bar 10). These findings demonstrate that *O*-GlcNAcylation of Sp1 is required for the inhibition of the HIV-1 LTR promoter by OGT.

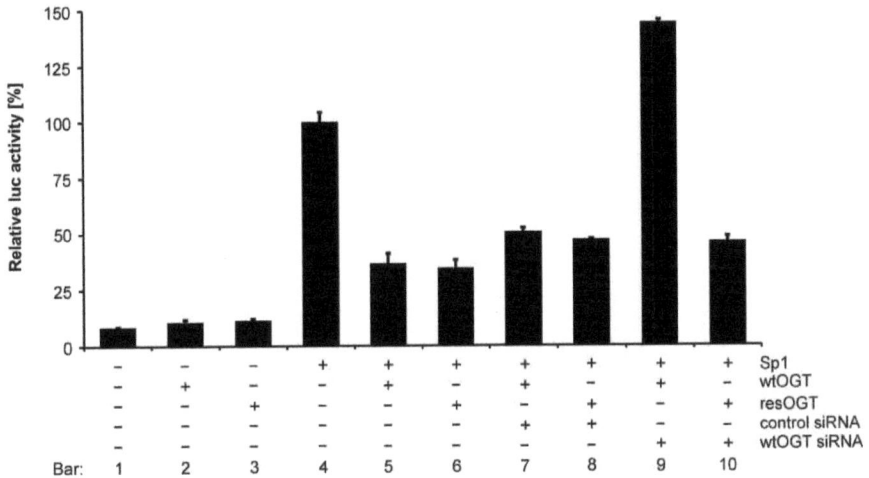

Fig. 28. Sp1 *O*-GlcNAcylation is necessary for the inhibition of the HIV-1 LTR. HEK 293T cells were transfected with the reporter construct LTR-Sp1$_{wt}$ along with plasmids encoding Sp1, wtOGT or resOGT. Total DNA amount was adjusted with pcDNA4. 10 nM control siRNA or siRNA targeting wtOGT were cotransfected and the luciferase activity was measured. The values were adjusted to the total amount of protein and indicate the means ± standard deviations from triplicate determinations. The results are presented in terms of percent of control (Sp1-induced LTR activity).

4.9 Increased OGT Expression Before Infection Increases OGT-Inhibitory Effect on HIV-1 Replication

In all previous experiments, the *O*-GlcNAcylation level was increased after infection with HIV-1 and after transfection of HIV-1 LTR encoding reporter plasmids. In order to analyze the effect of increased *O*-GlcNAcylation level prior to infection, Jurkat cells were stably transfected with OGT or d2EGFP, as control. Expression of recombinant OGT in selected clones was confirmed by RT-PCR with specific primers (Fig. 29A). The stably transfected cells were infected with VSV-G *env* pseudotyped HIV-1$_{NL4-3LucR-E-}$ and LTR activity was determined with luciferase assay. Of note, HIV-1 LTR activity was 40-fold lower in both OGT overexpressing Jurkat cell clones as compared to control cells expressing d2EGFP (Fig. 29B). These results demonstrate that OGT is a potent inhibitor of HIV-1 LTR activity.

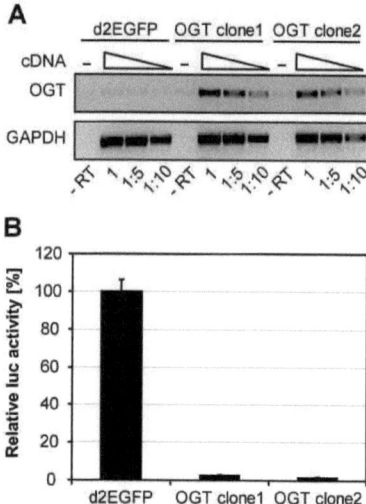

Fig. 29. OGT overexpression prior to infection amplifies the inhibitory effect on HIV-1 replication. (A) Jurkat cells were stably transfected with plasmids encoding either d2EGFP or OGT. Expression of recombinantly expressed OGT mRNA was confirmed by RT-PCR (upper panel) using specific primers. Amplification of GAPDH served as loading control (lower panel). Cellular cDNA was subjected to the amplification reactions in increasing dilutions: undiluted (1), 1:5 and 1:10. Control reactions were carried out in the absence of reverse transcriptase (-RT) with undiluted cDNA template. (B) Jurkat cells stably expressing d2EGFP or OGT were infected with VSV-G *env* pseudotyped HIV-1$_{NL4-3LucR-E-}$. Luciferase activity was measured 48 h after infection. The values indicate the means ± standard deviations from triplicate determinations.

5. Discussion

Regulation of gene expression by nutrients like glucose and GlcN is well established (68, 113, 160, 231). This demonstrates that cellular transcription adapts to environmental and metabolic changes. Viruses are unable to grow or reproduce outside a host cell. They hijack the cell's reproductive machinery in order to replicate and are therefore reliant on the environment of the host cell. Although viruses are able to create a favorable intracellular environment for their replication (107, 228, 256), it has also been shown, that they critically depend on the host cell metabolism (98) and that viral replication is affected by changes in the carbohydrate environment (207) and metabolism (93). The most prominent carbohydrate is the monosaccharide glucose. Two to three percent of the incoming glucose enter the hexosamine biosynthesis pathway and become metabolized to GlcNAc (238). UDP-GlcNAc is the substrate for O-GlcNAcylation mediated by OGT. This directed us to evaluate whether O-GlcNAc has an impact on HIV-1 replication.

Since Sp1 is critical and important for HIV-1 transcription (84), and Sp1 activity is modulated by O-GlcNAc (130), the effect of Sp1 O-GlcNAcylation on the activity of the HIV-1 LTR promoter was investigated. Luciferase experiments revealed that increased O-GlcNAc levels repress the HIV-1 LTR promoter activity. This required O-GlcNAcylation of Sp1 and the presence of Sp1-binding sites in the HIV-1 LTR.

5.1 O-GlcNAcylation Inhibits HIV-1 Transcription

GlcN is an amino sugar which bypasses the rate limiting reaction in the hexosamine biosynthesis pathway and becomes directly metabolized to UDP-GlcNAc. As the activity of OGT, the enzyme catalyzing O-GlcNAcylation, is primarily regulated by the availability of its substrate (156), treatment of cells with GlcN increases O-GlcNAcylation. The results presented in this work reveale that GlcN-treatment downregulates HIV-1 transcription in infected T cell lines and primary $CD4^+$ T lymphocytes. The inhibition correlated with increased levels of O-GlcNAcylated proteins, as detected by Western blot. This correlation was most evident in the results obtained with the primary $CD4^+$ T cells from two different donors (compare Fig. 14D to F): while donor 1 showed a highly significant downregulation in the HIV-1 transcription and a strong increase in O-GlcNAcylation upon GlcN-treatment, inhibition of HIV-1 transcription in donor 2 was less prominent. This correlated with a lower amount of O-GlcNAcylated proteins in the presence of GlcN.

Accordingly, overexpression of OGT in HIV-1 infected primary $CD4^+$ T lymphocytes inhibited HIV-1 transcription to the same extent as treatment with 16 mM GlcN (compare Fig. 17A with Fig.

Discussion

14D and E). Of note, the *O*-GlcNAcylation pattern induced by 16 mM GlcN was strongly similar to that one observed in cells overexpressing OGT (compare Fig. 14F and Fig. 17B). This indicates that both experimental setups lead to comparable levels of *O*-GlcNAcylation.

The effect of GlcN was further analyzed in HeLa cells stably expressing d1EGFP under the control of the HIV-1 LTR. These cells harbored the HIV-1 LTR promoter stably integrated into the cellular chromatin and thereby provided a suitable tool to investigate the effect of GlcN on the HIV-1 promoter activity, without the need to infect the cells. Thus, interference of GlcN with the nuclear transport or the integration of the viral genome could be excluded. GlcN inhibited the HIV-1 LTR-mediated gene transcription to a similar extent as in lymphocytes (compare Fig. 15), corroborating the assumption that GlcN directly represses the LTR promoter.

Importantly, stable overexpression of OGT strongly inhibited replication of HIV-1 in *de novo* infected lymphocytes (compare Fig. 29). OGT might therefore repress early steps in HIV-1 gene transcription, such as transcription from non-integrated DNA (216, 217), which is the earliest event following viral entry. Pre-integration transcription occurs in the absence of newly synthesized viral factors such as Tat (217) and relies on cellular factors like Sp1 and cyclin T1 (249). Altogether, these results suggest that increased *O*-GlcNAcylation prior to HIV-1 infection might efficiently prevent the onset of viral replication. OGT might therefore be a potent inhibitor of HIV-1 replication and may be involved in the regulation of the viral life cycle.

5.2 Sp1-Binding Sites and Sp1 Protein Are Necessary for the *O*-GlcNAc-Mediated Inhibition of HIV-1 Transcription

The inhibition of the HIV-1 LTR promoter appears to be mediated by the transcription factor Sp1 as supported by several lines of experimental evidence: first, it was demonstrated that *O*-GlcNAc-mediated inhibition of HIV-1 transcription required the presence of Sp1-binding sites in the HIV-1 LTR promoter (compare Fig. 18 and Fig. 19). The Sp1-binding sites are well conserved throughout the lentiviral HIV-1 LTRs (60) and mutation of one or more Sp1-binding sites in the basal HIV-1 LTR promoter leads to a strong delay in replication in human peripheral blood lymphocytes and in T cell lines (142, 173). This was also reflected in our observations that mutation of the Sp1-binding sites decreased HIV-1 LTR activity in transiently and stably transfected cells. However, Sp1-mutated HIV-1 LTR activity was still measurable and clearly inducible. Regardless, no significant *O*-GlcNAc-mediated inhibition of promoter activity was observed when the Sp1-binding sites were mutated.

Discussion

Detailed analyses of the individual Sp1-binding sites revealed that the promoter-proximal (Sp-I) and the promoter-medial (Sp-II) Sp1 motifs mediate the inhibitory effect of *O*-GlcNAc on the HIV-1 LTR promoter. Inactivation of the promoter-distal Sp1-binding site (Sp-III) in the HIV-1 LTR did not interfere with OGT-mediated inhibition (compare Fig. 20). This is in line with previously described observations that the two promoter-proximal Sp1 motifs are more important than the distal motif and sufficient to induce basal and Tat-mediated transcriptional activation (9).

Second, the presence of Sp1 protein was crucial for the inhibitory effect of *O*-GlcNAc on the HIV-1 LTR. Depletion of Sp1 expression by RNA interference completely abolished OGT-mediated inhibition of the HIV-1 LTR activity (compare Fig. 21). Interestingly, although Sp1 is implicated in the activation of a large number of genes, its depletion did not induce cytotoxicity in the time period and under the experimental conditions used here. This is in agreement with the observation of Philipsen and coworkers, who disrupted the mouse *Sp1* gene and found that Sp1-deficient embryonic stem cells were viable and showed normal growth characteristics under standard tissue-culture conditions (155). This might be attributed to the compensation of Sp1-triggered gene transcription by other Sp-family members, including Sp2, Sp3 and Sp4. Like Sp1, all three proteins contain zinc fingers and glutamine- and serine-rich motifs and are ubiquitously expressed (77, 123, 158). Concerning the HIV-1 LTR, Sp1 and Sp4 have been shown to activate LTR-driven expression, whereas Sp3 acts as a repressor (150, 158). However, Sp3 has also the potential to activate transcription of certain otherwise Sp1 regulated promoters (221). Despite not being found crucial for cell survival, Sp1 played a key role in the inhibition of the HIV-1 LTR by OGT and its depletion could not be substituted by other members of the Sp-family.

5.3 *O*-GlcNAcylation of Sp1 Inhibits the HIV-1 LTR in a Dominant Manner

The HIV-1 LTR promoter containing only the three Sp1-binding sites (LTR-Sp1$_{wt}$) was inhibited by OGT overexpression (compare Fig. 18), indicating that *O*-GlcNAcylation of Sp1 inhibits transcription from the LTR-Sp1$_{wt}$ promoter. In contrast, the HIV-1 LTR promoter containing only functional NF-κB-binding sites but mutated Sp1-binding sites (LTR-κB-Sp1$_{mut}$) was activated by OGT overexpression, indicating that *O*-GlcNAcylation of NF-κB increases the transcription rate from the LTR-κB-Sp1$_{mut}$ promoter. In fact, *O*-GlcNAcylation of NF-κB has been described and has been suggested to enhance the nuclear translocation of NF-κB (69, 103). However, the transcription rate of the wild type HIV-1 LTR (LTR-κB-Sp1$_{wt}$) containing both NF-κB- and Sp1-binding sites was inhibited by *O*-GlcNAcylation of Sp1. This suggests that *O*-GlcNAcylation of Sp1 inhibits the activity of the HIV-1 LTR in a dominant manner. Accordingly, unlike the NF-κB-binding sites, the

Sp1-binding sites are highly conserved throughout primate lentiviral LTRs (60, 249) and the HIV-1 subtypes A through G (104). This implies that the regulation of Sp1-mediated gene transcription for lentiviral LTRs is also well conserved and presumably dominant over NF-κB regulation.

Moreover, *O*-GlcNAcylated Sp1 seems to act as a transcriptional repressor at the HIV-1 LTR. Several findings support this hypothesis: first, *O*-GlcNAcylation did not alter the DNA binding affinity of Sp1 (compare Fig. 25), providing evidence that *O*-GlcNAcylated Sp1 binds to the HIV-1 LTR promoter. Second, mutations of the Sp1 binding sites in the HIV-1 LTR promoter abolished the inhibitory effect of OGT on the HIV-1 gene expression. Third, depletion of Sp1 protein completely eliminated OGT-mediated inhibition of the LTR promoter activity. The latter two findings argue that *O*-GlcNAcylated Sp1 must be present and bound to the HIV-1 LTR promoter in order to mediate the inhibition of the promoter activity by OGT. Thus, *O*-GlcNAcylation converts the transcriptional activator Sp1 into a transcriptional repressor.

5.4 *O*-GlcNAcylation of Sp1 Specifically Inhibits HIV-1 LTR Activity

Sp1 was found to be *O*-GlcNAcylated by OGT (compare Fig. 22) leading to a decreased ability to activate the HIV-1 LTR promoter. These findings are in accordance with the results of Kudlow and colleagues, who showed that *O*-GlcNAcylation of Sp1 decreases its capability to activate an artificial promoter containing tandem GC-boxes (247). However, it has to be emphasized that *O*-GlcNAcylation of Sp1 does not generally inhibit transcription. The human EF1α promoter, which also contains Sp1 binding sites (166, 229, 233), is activated by *O*-GlcNAcylation of Sp1 (compare Fig. 23). Additionally, others have shown that the expression of plasminogen activator inhibitor-1 (45, 67, 68), calmodulin (118), and argininosuccinate synthetase (15, 16) is also increased by Sp1 *O*-GlcNAcylation. Thus, *O*-GlcNAcylation of Sp1 differentially modulates gene expression.

Sp1 is a ubiquitous transcription factor regulating the expression of several hundreds of proteins. Hence, it is not surprising that posttranslational modifications modulate the specific activity of Sp1 on different promoters. Importantly, phosphorylation of different residues on Sp1 results in distinct Sp1 activity (25), indicating that both, the type and the site of modification are decisive. This is exemplified by the fact that phosphorylation of Sp1 at Thr661 (originally quoted as Thr579) decreases the transcriptional activity of Sp1, while phosphorylation at Ser52 (originally quoted as Ser59) increases Sp1 activity (3, 25, 58). *O*-GlcNAcylation appears to have a yin-yang relationship with phosphorylation (33, 87). Therefore, *O*-GlcNAcylation may have a similar potential to differentially regulate the activity of Sp1. This is well in agreement with the fact that Sp1 bears at least eight *O*-GlcNAc sites (101). Hence, it is conceivable that *O*-GlcNAcylation of Sp1 at a specific site increases, while *O*-GlcNAcylation of Sp1 at another site decreases its transactivation ability. Furthermore, *O*-GlcNAc competes with *O*-phosphate for the same serine and threonine residues (86). Phosphorylation of Sp1 at Ser213 (originally quoted as Ser131) has been described to increase the transcriptional activity of the HIV-1 LTR (26). In this context, it is possible that under increased *O*-GlcNAc conditions, *O*-GlcNAcylation of Sp1 at this site or at adjacent sites might occur, thereby preventing phosphorylation and leading to a decreased activity at the HIV-1 LTR. In contrast, *O*-GlcNAc modification at this site(s) of Sp1 might have no impact on the regulation of the EF1α control promoter used in this study. The possible mechanisms explaining how *O*-GlcNAcylation alters the transcriptional activity of Sp1 will be discussed in the next chapter.

5.5 Mechanism for *O*-GlcNAc-Mediated Inhibition of HIV-1 Transcription

The transcriptional activity of Sp1 is regulated by interactions with other proteins (6, 51, 54, 112, 138, 241). Similar to phosphorylation (25), *O*-GlcNAcylation of Sp1 might result in different conformations, altering the ability of Sp1 to interact with other proteins. Two possible mechanisms might explain how *O*-GlcNAcylation inhibits the transcriptional activity of Sp1. First, *O*-GlcNAcylation of Sp1 might block its interaction with factors of the basal transcription machinery. This hypothesis is enforced by the observation that *O*-GlcNAcylation of Sp1 disrupts its interaction with the *Drosophila* transcriptional coactivator $TAF_{II}110$ (191, 247). In this regard, it is conspicuous that the Sp1 activation domains A and/or B are essential for interactions with basal transcription factors (241) (Fig. 30A). Interestingly, it has recently been shown that the majority of the Sp1 *O*-GlcNAcylation sites reside within the second serine/threonine-rich region (122), located between the activation domains A and B. This was also confirmed by analysis of *O*-GlcNAc attachment sites in Sp1 using the *YinOYang1.2 Prediction Server* (75, 76): more than 40% of the predicted *O*-GlcNAc sites reside in the region between the amino acids 264 and 344 (Fig. 30B).

Discussion

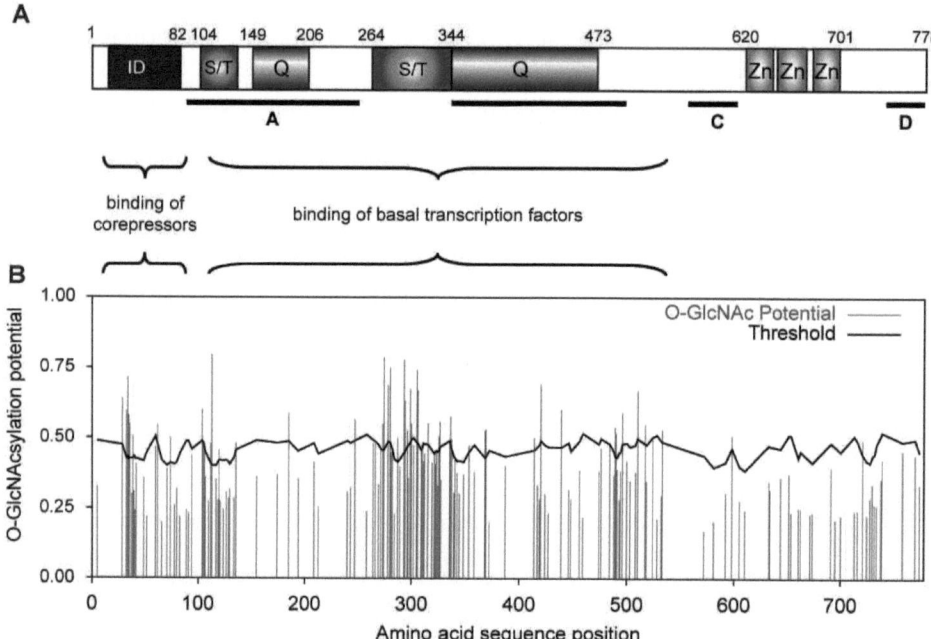

Fig. 30. Functional domains of Sp1 and *O*-GlcNAc sites in the Sp1 sequence. (A) Sp1 has 778 amino acids (aa). Numbers above the Sp1 structure indicate the aa position in the protein. ID, inhibitory domain; S/T, serine/threonine-rich regions; Q, glutamine-rich regions; Zn, zinc finger DNA-binding domain. The transactivation domains A, B, C and D are indicated by the thick black bars. The domains necessary for the interaction with corepressors or basal transcription factors are indicated by colons. (B) Predicted sites of Sp1 *O*-GlcNAcylation at serine or threonine residues by the *YinOYang 1.2 Prediction Server* (http://www.cbs.dtu.dk/services/YinOYang/). The x-axis represents the sequence from N-terminal to C-terminal ends. The aa position is indicated at the bottom. Vertical impulses (grey) indicate the *O*-GlcNAc potentials. Potentials crossing the threshold (black wavy horizontal) line represent predicted *O*-GlcNAcylated sites.

The second mechanism might rely on the ability of Sp1 to recruit corepressors to the promoter, such as NCor, BCor, SMRT (138), HDACs (42), and mSin3a (258). The inhibitory domain is required for most of these interactions (Fig. 30A). *O*-GlcNAcylation of Sp1 in this region might therefore lead to conformational changes, enabling it to interact with corepressors. This is concordant with the several predicted *O*-GlcNAc attachment sites in the inhibitory domain of Sp1 (Fig. 30B).

However, these mechanisms might be valid and apply for all Sp1-regulated promoters. In contrast, the experiments described above prove that *O*-GlcNAcylation of Sp1 does not necessarily inhibit promoter activity. In the experimental setup presented in this work, only the HIV-1 LTR promoter activity was inhibited, while the Sp1-regulated activity of the EF1α promoter was increased. The differential effects of Sp1-*O*-GlcNAcylation might be modulated by the individual composition of the Sp1-containing transcription complexes at diverse promoters. Thus, one can postulate that *O*-GlcNAcylation of Sp1 selectively inhibits an LTR specific mechanism. Regarding the two possible mechanisms described above, *O*-GlcNAcylation might either inhibit the interaction of Sp1 with transcription factors specifically targeted to the HIV-1 LTR, or enable Sp1-dependent recruitment of corepressors to the viral promoter. The gene expression from the HIV-1 LTR is regulated by a transcription factor complex composed of basal/general transcription factors (such as TFIID, TBP, and RNA polymerase II), specific transcription factors (like Sp1 and NF-κB, the positive transcription elongation factor [P-TEFb] composed of cyclin T1 and cyclin dependent kinase 9), and the viral transactivator Tat (246). Direct interactions of Sp1 with NF-κB (179), cyclin T1 (249), and Tat (111) have been described and exhibit possible areas of attack upon *O*-GlcNAcylation of Sp1. For example, it has been shown that direct interaction between DNA-bound Sp1 and cyclin T1 can robustly activate the HIV-1 LTR promoter (249). This interaction can occur in the absence or presence of Tat. Strikingly, impaired interaction between DNA-bound Sp1 and cyclin T1 would inhibit not only the HIV-1 transcription from integrated proviral DNA, but also transcription from non-integrated viral DNA and might explain the strong reduction in HIV-1 gene expression observed when OGT was overexpressed prior to infection (compare Fig. 29). Alternatively, *O*-GlcNAcylation of Sp1 might lead to the recruitment of a repressor complex composed of c-Myc and HDAC1 to the HIV-1 LTR, which has been shown to establish proviral latency (105). It remains to be determined in further experiments which of the two options above apply for the OGT-mediated inhibition of the HIV-1 LTR.

In order to allow transcriptional activation to occur, *O*-GlcNAc modifications have to be removed at the promoter. This is in agreement with our observations that *O*-GlcNAcylation of Sp1 does neither decrease its nuclear concentration (compare Fig. 24), nor its DNA binding affinity (compareFig. 25). Furthermore, OGT and NCOAT were described to associate with proteins at the chromatin (130). OGT can be recruited to promoters *via* corepressors such as mSin3A, thereby repressing transcription (248). Accordingly, NCOAT, which has both *O*-GlcNAcase activity to remove inhibitory sugars and HAT activity to acetylate histones, enables activation of eukaryotic gene transcription (225). Moreover, both enzymes associate in one complex (240), thereby allowing a dynamic regulation of HIV-1 gene expression in response to environmental stimuli.

5.6 Role of Sp1-*O*-GlcNAcylation in the Immune System in the Context of Viral Infection

Many reports have evaluated the effect of highly active antiretroviral therapy (HAART) on glucose metabolism (46, 50, 149) and the appearance of diabetes as a consequence of HAART (126, 157).

One product strongly coupled to the metabolic state of the cells is UDP-GlcNAc, the substrate for OGT-mediated *O*-GlcNAcylation. So far, no studies have investigated the effect of glucose homeostasis on HIV-1 replication and only few reports exist, assigning *O*-GlcNAc a role in the immune system: Over 15 years ago, Kearse and Hart observed a rapid alteration in the *O*-GlcNAc levels on nuclear and cytoplasmic proteins after stimulation of T cells with mitogens such as Concanavalin A and Phorbol 12-myristate 13-acetate/ionomycin and suggested that *O*-GlcNAc modification may play an important role in the early stages of T lymphocyte activation (116). Since then, the progress in *O*-GlcNAc signaling research was hampered due to the limited availability of tools to detect *O*-GlcNAc-modified proteins. About one decade later, it was described that the activity of the p65 subunit of NF-κB is modulated by *O*-GlcNAc (103). This was just recently confirmed by Guerini and coworkers, who further reported that NFAT, another important lymphocyte transcription factor, is also regulated by the transient modification with *O*-GlcNAc (69). In the same work, it has been shown that OGT is crucial for efficient activation of T lymphocytes, as knockdown of OGT by siRNA blocked T cell activation. Thus, in line with the high expression levels of OGT in T lymphocytes, glucose homeostasis and *O*-GlcNAc metabolism may be key regulatory mechanisms for T cell activation.

Furthermore, it has been observed that compounds like GlcN and 2-DG have antiviral effects on enveloped RNA viruses, such as Sindbis virus, Semliki Forest virus, Influenza virus, VSV and Newcastle Disease virus (110, 124, 201). The antiviral effects of GlcN and 2-DG have been attributed to two different mechanisms: blocked synthesis of viral glycoproteins and decreased viral RNA synthesis without impairing RNA and protein synthesis in non-infected cells (202). The results presented in this work prove for the first time, that GlcN can directly block viral promoter activity *via O*-GlcNAcylation of the transcription factor Sp1. This is mediated by the enzymatic activity of OGT. *O*-GlcNAcylation of Sp1 after infection reduces HIV-1 gene transcription to about 50%, while *O*-GlcNAcylation prior to infection strongly prevents the onset of viral replication. Thus, in the context of HIV-1 infection, *O*-GlcNAcylated Sp1 might represent an antiviral cellular response. Whether this is also true for other viruses, especially enveloped RNA viruses, remains to be determined in further studies.

5.7 Possible Implications of Sp1-*O*-GlcNAcylation in the Antiretroviral Therapy

The results presented in this study indicate that the *O*-GlcNAc level, and thus, the glucose metabolism, may influence the life cycle of HIV-1. Accordingly, inducers of Sp1 *O*-GlcNAcylation such as GlcN and 2-DG, which are approved in clinical treatment of osteoarthritis and human genital herpes infections, respectively (11, 183), may support HAART. Interestingly, in the treatment of osteoarthritis, GlcN is applied in a daily dose of 1,500 mg/day (188), corresponding to 1.16 mM GlcN, which is in the range of the concentrations used in this study (0.25 – 16 mM GlcN). Apart from its application in the treatment of human genital herpes virus, 2-DG has the ability to inhibit glycolysis and thereby the proliferation of fast-growing cells and is therefore tested in clinical trials I/II for the treatment of advanced solid tumors and prostate cancer (http://clinicaltrials.gov). Thus, GlcN and 2-DG might be promising antiretroviral compounds. Hence, the establishment of a metabolic treatment may supplement the repertoire of antiretroviral therapies against AIDS.

6. References

1. **Abdelrahim, M., and S. Safe.** 2005. Cyclooxygenase-2 Inhibitors Decrease Vascular Endothelial Growth Factor Expression in Colon Cancer Cells by Enhanced Degradation of Sp1 and Sp4 Proteins. Mol Pharmacol **68**:317-329.

2. **Adachi, A., H. E. Gendelman, S. Koenig, T. Folks, R. Willey, A. Rabson, and M. A. Martin.** 1986. Production of acquired immunodeficiency syndrome-associated retrovirus in human and nonhuman cells transfected with an infectious molecular clone. J Virol **59**:284-91.

3. **Armstrong, S. A., D. A. Barry, R. W. Leggett, and C. R. Mueller.** 1997. Casein kinase II-mediated phosphorylation of the C terminus of Sp1 decreases its DNA binding activity. J Biol Chem **272**:13489-95.

4. **Bailey, J. R., A. R. Sedaghat, T. Kieffer, T. Brennan, P. K. Lee, M. Wind-Rotolo, C. M. Haggerty, A. R. Kamireddi, Y. Liu, J. Lee, D. Persaud, J. E. Gallant, J. Cofrancesco, Jr., T. C. Quinn, C. O. Wilke, S. C. Ray, J. D. Siliciano, R. E. Nettles, and R. F. Siliciano.** 2006. Residual human immunodeficiency virus type 1 viremia in some patients on antiretroviral therapy is dominated by a small number of invariant clones rarely found in circulating CD4+ T cells. J Virol **80**:6441-57.

5. **Ball, L. E., M. N. Berkaw, and M. G. Buse.** 2006. Identification of the major site of O-linked beta-N-acetylglucosamine modification in the C terminus of insulin receptor substrate-1. Mol Cell Proteomics **5**:313-23.

6. **Bargonetti, J., A. Chicas, D. White, and C. Prives.** 1997. p53 represses Sp1 DNA binding and HIV-LTR directed transcription. Cell Mol Biol (Noisy-le-grand) **43**:935-49.

7. **Baron, A. D., J. S. Zhu, J. H. Zhu, H. Weldon, L. Maianu, and W. T. Garvey.** 1995. Glucosamine induces insulin resistance in vivo by affecting GLUT 4 translocation in skeletal muscle. Implications for glucose toxicity. J Clin Invest **96**:2792-801.

8. **Barre-Sinoussi, F., J. C. Chermann, F. Rey, M. T. Nugeyre, S. Chamaret, J. Gruest, C. Dauguet, C. Axler-Blin, F. Vezinet-Brun, C. Rouzioux, W. Rozenbaum, and L. Montagnier.** 1983. Isolation of a T-lymphotropic retrovirus from a patient at risk for acquired immune deficiency syndrome (AIDS). Science **220**:868-71.

9. **Berkhout, B., and K. T. Jeang.** 1992. Functional roles for the TATA promoter and enhancers in basal and Tat-induced expression of the human immunodeficiency virus type 1 long terminal repeat. J Virol **66**:139-49.

10. **Billon, N., D. Carlisi, M. B. Datto, L. A. van Grunsven, A. Watt, X. F. Wang, and B. B. Rudkin.** 1999. Cooperation of Sp1 and p300 in the induction of the CDK inhibitor p21WAF1/CIP1 during NGF-mediated neuronal differentiation. Oncogene **18**:2872-82.

11. **Blough, H. A., and R. L. Giuntoli.** 1979. Successful treatment of human genital herpes infections with 2-deoxy-D-glucose. JAMA **241**:2798-2801.

12. **Bonehill, A., C. Heirman, S. Tuyaerts, A. Michiels, K. Breckpot, F. Brasseur, Y. Zhang, P. Van Der Bruggen, and K. Thielemans.** 2004. Messenger RNA-electroporated dendritic cells presenting MAGE-A3 simultaneously in HLA class I and class II molecules. J Immunol **172**:6649-57.

13. **Bowe, D. B., A. Sadlonova, C. A. Toleman, Z. Novak, Y. Hu, P. Huang, S. Mukherjee, T. Whitsett, A. R. Frost, A. J. Paterson, and J. E. Kudlow.** 2006. O-GlcNAc integrates the proteasome and transcriptome to regulate nuclear hormone receptors. Mol Cell Biol **26**:8539-50.

14. **Brady, H. J., D. J. Pennington, C. G. Miles, and E. A. Dzierzak.** 1993. CD4 cell surface downregulation in HIV-1 Nef transgenic mice is a consequence of intracellular sequestration. EMBO J **12**:4923-32.

15. **Brasse-Lagnel, C., A. Fairand, A. Lavoinne, and A. Husson.** 2003. Glutamine stimulates argininosuccinate synthetase gene expression through cytosolic O-glycosylation of Sp1 in Caco-2 cells. J Biol Chem **278**:52504-10.

16. **Brasse-Lagnel, C., A. Lavoinne, D. Loeber, A. Fairand, C. Bole-Feysot, N. Deniel, and A. Husson.** 2007.

Glutamine and interleukin-1beta interact at the level of Sp1 and nuclear factor-kappaB to regulate argininosuccinate synthetase gene expression. FEBS J **274**:5250-62.

17. **Briggs, M. R., J. T. Kadonaga, S. P. Bell, and R. Tjian.** 1986. Purification and biochemical characterization of the promoter-specific transcription factor, Sp1. Science **234**:47-52.

18. **Buse, M. G., K. A. Robinson, B. A. Marshall, R. C. Hresko, and M. M. Mueckler.** 2002. Enhanced O-GlcNAc protein modification is associated with insulin resistance in GLUT1-overexpressing muscles. Am J Physiol Endocrinol Metab **283**:E241-50.

19. **Calman, A. F., M. P. Busch, G. N. Vyas, T. M. McHugh, D. P. Stites, and B. M. Peterlin.** 1988. Transcription and replication of human immunodeficiency virus-1 in B lymphocytes in vitro. AIDS **2**:185-93.

20. **Chakraborty, A., D. Saha, A. Bose, M. Chatterjee, and N. K. Gupta.** 1994. Regulation of eIF-2 alpha-subunit phosphorylation in reticulocyte lysate. Biochemistry **33**:6700-6.

21. **Chan, D. C., D. Fass, J. M. Berger, and P. S. Kim.** 1997. Core structure of gp41 from the HIV envelope glycoprotein. Cell **89**:263-73.

22. **Chang, L. J., V. Urlacher, T. Iwakuma, Y. Cui, and J. Zucali.** 1999. Efficacy and safety analyses of a recombinant human immunodeficiency virus type 1 derived vector system. Gene Ther **6**:715-28.

23. **Cheng, X., and G. W. Hart.** 2001. Alternative O-Glycosylation/O-Phosphorylation of Serine-16 in Murine Estrogen Receptor beta. POST-TRANSLATIONAL REGULATION OF TURNOVER AND TRANSACTIVATION ACTIVITY. J. Biol. Chem. **276**:10570-10575.

24. **Chou, T.-Y., G. W. Hart, and C. V. Dang.** 1995. c-Myc Is Glycosylated at Threonine 58, a Known Phosphorylation Site and a Mutational Hot Spot in Lymphomas. J. Biol. Chem. **270**:18961-18965.

25. **Chu, S., and T. J. Ferro.** 2005. Sp1: Regulation of gene expression by phosphorylation. Gene **348**:1-11.

26. **Chun, R. F., O. J. Semmes, C. Neuveut, and K. T. Jeang.** 1998. Modulation of Sp1 phosphorylation by human immunodeficiency virus type 1 Tat. J Virol **72**:2615-29.

27. **Chun, T. W., L. Carruth, D. Finzi, X. Shen, J. A. DiGiuseppe, H. Taylor, M. Hermankova, K. Chadwick, J. Margolick, T. C. Quinn, Y. H. Kuo, R. Brookmeyer, M. A. Zeiger, P. Barditch-Crovo, and R. F. Siliciano.** 1997. Quantification of latent tissue reservoirs and total body viral load in HIV-1 infection. Nature **387**:183-8.

28. **Chun, T. W., R. T. Davey, Jr., D. Engel, H. C. Lane, and A. S. Fauci.** 1999. Re-emergence of HIV after stopping therapy. Nature **401**:874-5.

29. **Chun, T. W., D. Finzi, J. Margolick, K. Chadwick, D. Schwartz, and R. F. Siliciano.** 1995. In vivo fate of HIV-1-infected T cells: quantitative analysis of the transition to stable latency. Nat Med **1**:1284-90.

30. **Chun, T. W., L. Stuyver, S. B. Mizell, L. A. Ehler, J. A. Mican, M. Baseler, A. L. Lloyd, M. A. Nowak, and A. S. Fauci.** 1997. Presence of an inducible HIV-1 latent reservoir during highly active antiretroviral therapy. Proc Natl Acad Sci U S A **94**:13193-7.

31. **Cole, R. N., and G. W. Hart.** 1999. Glycosylation sites flank phosphorylation sites on synapsin I: O-linked N-acetylglucosamine residues are localized within domains mediating synapsin I interactions. J Neurochem **73**:418-28.

32. **Comer, F. I., and G. W. Hart.** 1999. O-GlcNAc and the control of gene expression. Biochim Biophys Acta **1473**:161-71.

33. **Comer, F. I., and G. W. Hart.** 2001. Reciprocity between O-GlcNAc and O-phosphate on the carboxyl terminal domain of RNA polymerase II. Biochemistry **40**:7845-52.

34. **Connor, R. I., B. K. Chen, S. Choe, and N. R. Landau.** 1995. Vpr Is Required for Efficient Replication of Human Immunodeficiency Virus Type-1 in Mononuclear Phagocytes. Virology **206**:935-944.

References

35. **Courey, A. J., D. A. Holtzman, S. P. Jackson, and R. Tjian.** 1989. Synergistic activation by the glutamine-rich domains of human transcription factor Sp1. Cell **59**:827-36.

36. **Courey, A. J., and R. Tjian.** 1988. Analysis of Sp1 in vivo reveals multiple transcriptional domains, including a novel glutamine-rich activation motif. Cell **55**:887-98.

37. **Darke, P. L., R. F. Nutt, S. F. Brady, V. M. Garsky, T. M. Ciccarone, C. T. Leu, P. K. Lumma, R. M. Freidinger, D. F. Veber, and I. S. Sigal.** 1988. HIV-1 protease specificity of peptide cleavage is sufficient for processing of gag and pol polyproteins. Biochem Biophys Res Commun **156**:297-303.

38. **Datta, B., M. K. Ray, D. Chakrabarti, and N. K. Gupta.** 1988. Roles of eIF-2 and eIF-2-associated proteins in regulation of protein synthesis during growth of animal cells in culture. Indian J Biochem Biophys **25**:478-82.

39. **Datta, B., M. K. Ray, D. Chakrabarti, D. E. Wylie, and N. K. Gupta.** 1989. Glycosylation of eukaryotic peptide chain initiation factor 2 (eIF-2)-associated 67-kDa polypeptide (p67) and its possible role in the inhibition of eIF-2 kinase-catalyzed phosphorylation of the eIF-2 alpha-subunit. J Biol Chem **264**:20620-4.

40. **Davey, R. T., Jr., N. Bhat, C. Yoder, T. W. Chun, J. A. Metcalf, R. Dewar, V. Natarajan, R. A. Lempicki, J. W. Adelsberger, K. D. Miller, J. A. Kovacs, M. A. Polis, R. E. Walker, J. Falloon, H. Masur, D. Gee, M. Baseler, D. S. Dimitrov, A. S. Fauci, and H. C. Lane.** 1999. HIV-1 and T cell dynamics after interruption of highly active antiretroviral therapy (HAART) in patients with a history of sustained viral suppression. Proc Natl Acad Sci U S A **96**:15109-14.

41. **Deng, H., R. Liu, W. Ellmeier, S. Choe, D. Unutmaz, M. Burkhart, P. Di Marzio, S. Marmon, R. E. Sutton, C. M. Hill, C. B. Davis, S. C. Peiper, T. J. Schall, D. R. Littman, and N. R. Landau.** 1996. Identification of a major co-receptor for primary isolates of HIV-1. Nature **381**:661-6.

42. **Doetzlhofer, A., H. Rotheneder, G. Lagger, M. Koranda, V. Kurtev, G. Brosch, E. Wintersberger, and C. Seiser.** 1999. Histone deacetylase 1 can repress transcription by binding to Sp1. Mol Cell Biol **19**:5504-11.

43. **Dong, D. L., and G. W. Hart.** 1994. Purification and characterization of an O-GlcNAc selective N-acetyl-beta-D-glucosaminidase from rat spleen cytosol. J Biol Chem **269**:19321-30.

44. **Drosopoulos, W. C., L. F. Rezende, M. A. Wainberg, and V. R. Prasad.** 1998. Virtues of being faithful: can we limit the genetic variation in human immunodeficiency virus? J Mol Med **76**:604-12.

45. **Du, X. L., D. Edelstein, L. Rossetti, I. G. Fantus, H. Goldberg, F. Ziyadeh, J. Wu, and M. Brownlee.** 2000. Hyperglycemia-induced mitochondrial superoxide overproduction activates the hexosamine pathway and induces plasminogen activator inhibitor-1 expression by increasing Sp1 glycosylation. Proc Natl Acad Sci U S A **97**:12222-6.

46. **Dube, M. P.** 2000. Disorders of glucose metabolism in patients infected with human immunodeficiency virus. Clin Infect Dis **31**:1467-75.

47. **DuBridge, R. B., P. Tang, H. C. Hsia, P. M. Leong, J. H. Miller, and M. P. Calos.** 1987. Analysis of mutation in human cells by using an Epstein-Barr virus shuttle system. Mol. Cell. Biol. **7**:379-387.

48. **Dynan, W. S., and R. Tjian.** 1983. Isolation of transcription factors that discriminate between different promoters recognized by RNA polymerase II. Cell **32**:669-680.

49. **Dynan, W. S., and R. Tjian.** 1983. The promoter-specific transcription factor Sp1 binds to upstream sequences in the SV40 early promoter. Cell **35**:79-87.

50. **El-Sadr, W. M., C. M. Mullin, A. Carr, C. Gibert, C. Rappoport, F. Visnegarwala, C. Grunfeld, and S. S. Raghavan.** 2005. Effects of HIV disease on lipid, glucose and insulin levels: results from a large antiretroviral-naïve cohort. HIV Medicine **6**:114-121.

51. **Emami, K. H., T. W. Burke, and S. T. Smale.** 1998. Sp1 activation of a TATA-less promoter requires a species-specific interaction involving transcription factor IID. Nucleic Acids Res **26**:839-46.

52. **Emerman, M.** 1996. HIV-1, Vpr and the cell cycle. Curr Biol **6**:1096-103.

References

53. **Emili, A., J. Greenblatt, and C. J. Ingles.** 1994. Species-specific interaction of the glutamine-rich activation domains of Sp1 with the TATA box-binding protein. Mol Cell Biol **14:**1582-93.

54. **Enya, K., H. Hayashi, T. Takii, N. Ohoka, S. Kanata, T. Okamoto, and K. Onozaki.** 2008. The interaction with Sp1 and reduction in the activity of histone deacetylase 1 are critical for the constitutive gene expression of IL-1{alpha} in human melanoma cells. J Leukoc Biol **83:**190-9.

55. **Featherstone, C., M. K. Darby, and L. Gerace.** 1988. A monoclonal antibody against the nuclear pore complex inhibits nucleocytoplasmic transport of protein and RNA in vivo. J Cell Biol **107:**1289-97.

56. **Feng, Y., C. C. Broder, P. E. Kennedy, and E. A. Berger.** 1996. HIV-1 entry cofactor: functional cDNA cloning of a seven-transmembrane, G protein-coupled receptor. Science **272:**872-7.

57. **Finzi, D., M. Hermankova, T. Pierson, L. M. Carruth, C. Buck, R. E. Chaisson, T. C. Quinn, K. Chadwick, J. Margolick, R. Brookmeyer, J. Gallant, M. Markowitz, D. D. Ho, D. D. Richman, and R. F. Siliciano.** 1997. Identification of a reservoir for HIV-1 in patients on highly active antiretroviral therapy. Science **278:**1295-300.

58. **Fojas de Borja, P., N. K. Collins, P. Du, J. Azizkhan-Clifford, and M. Mudryj.** 2001. Cyclin A-CDK phosphorylates Sp1 and enhances Sp1-mediated transcription. Embo J **20:**5737-47.

59. **Frazer, I. H., I. R. Mackay, R. M. Crapper, B. Jones, I. D. Gust, M. G. Sarngadharan, D. C. Campbell, and B. Ungar.** 1986. Immunological abnormalities in asymptomatic homosexual men: correlation with antibody to HTLV-III and sequential changes over two years. Q J Med **61:**921-33.

60. **Frech, K., R. Brack-Werner, and T. Werner.** 1996. Common Modular Structure of Lentivirus LTRs. Virology **224:**256-267.

61. **Gao, F., E. Bailes, D. L. Robertson, Y. Chen, C. M. Rodenburg, S. F. Michael, L. B. Cummins, L. O. Arthur, M. Peeters, G. M. Shaw, P. M. Sharp, and B. H. Hahn.** 1999. Origin of HIV-1 in the chimpanzee Pan troglodytes troglodytes. Nature **397:**436-41.

62. **Gao, Y., L. Wells, F. I. Comer, G. J. Parker, and G. W. Hart.** 2001. Dynamic O-glycosylation of nuclear and cytosolic proteins: cloning and characterization of a neutral, cytosolic beta-N-acetylglucosaminidase from human brain. J Biol Chem **276:**9838-45.

63. **Gaynor, R.** 1992. Cellular transcription factors involved in the regulation of HIV-1 gene expression. AIDS **6:**347-63.

64. **Gheysen, D., E. Jacobs, F. de Foresta, C. Thiriart, M. Francotte, D. Thines, and M. De Wilde.** 1989. Assembly and release of HIV-1 precursor Pr55gag virus-like particles from recombinant baculovirus-infected insect cells. Cell **59:**103-12.

65. **Gidoni, D., J. T. Kadonaga, H. Barrera-Saldana, K. Takahashi, P. Chambon, and R. Tjian.** 1985. Bidirectional SV40 transcription mediated by tandem Sp1 binding interactions. Science **230:**511-7.

66. **Gill, G., E. Pascal, Z. H. Tseng, and R. Tjian.** 1994. A glutamine-rich hydrophobic patch in transcription factor Sp1 contacts the dTAFII110 component of the Drosophila TFIID complex and mediates transcriptional activation. Proc Natl Acad Sci U S A **91:**192-6.

67. **Goldberg, H. J., C. I. Whiteside, and I. G. Fantus.** 2002. The hexosamine pathway regulates the plasminogen activator inhibitor-1 gene promoter and Sp1 transcriptional activation through protein kinase C-beta I and - delta. J Biol Chem **277:**33833-41.

68. **Goldberg, H. J., C. I. Whiteside, G. W. Hart, and I. G. Fantus.** 2006. Posttranslational, reversible O-glycosylation is stimulated by high glucose and mediates plasminogen activator inhibitor-1 gene expression and Sp1 transcriptional activity in glomerular mesangial cells. Endocrinology **147:**222-31.

69. **Golks, A., T. T. Tran, J. F. Goetschy, and D. Guerini.** 2007. Requirement for O-linked N-acetylglucosaminyltransferase in lymphocytes activation. EMBO J **26:**4368-79.

70. **Gomez-Gonzalo, M., M. Carretero, J. Rullas, E. Lara-Pezzi, J. Aramburu, B. Berkhout, J. Alcami, and**

References

M. Lopez-Cabrera. 2001. The hepatitis B virus X protein induces HIV-1 replication and transcription in synergy with T-cell activation signals: functional roles of NF-kappaB/NF-AT and SP1-binding sites in the HIV-1 long terminal repeat promoter. J Biol Chem **276**:35435-43.

71. Gottlieb, M. S., R. Schroff, H. M. Schanker, J. D. Weisman, P. T. Fan, R. A. Wolf, and A. Saxon. 1981. Pneumocystis carinii pneumonia and mucosal candidiasis in previously healthy homosexual men: evidence of a new acquired cellular immunodeficiency. N Engl J Med **305**:1425-31.

72. Greenberg, M. E., A. J. Iafrate, and J. Skowronski. 1998. The SH3 domain-binding surface and an acidic motif in HIV-1 Nef regulate trafficking of class I MHC complexes. EMBO J **17**:2777-89.

73. Griffith, L. S., and B. Schmitz. 1999. O-linked N-acetylglucosamine levels in cerebellar neurons respond reciprocally to pertubations of phosphorylation. Eur J Biochem **262**:824-31.

74. Guinez, C., W. Morelle, J. C. Michalski, and T. Lefebvre. 2005. O-GlcNAc glycosylation: a signal for the nuclear transport of cytosolic proteins? Int J Biochem Cell Biol **37**:765-74.

75. Gupta, R., H. Birch, K. Rapacki, S. Brunak, and J. E. Hansen. 1999. O-GLYCBASE version 4.0: a revised database of O-glycosylated proteins. Nucleic Acids Res **27**:370-2.

76. Gupta, R., and S. Brunak. 2002. Prediction of glycosylation across the human proteome and the correlation to protein function. Pac Symp Biocomput:310-22.

77. Hagen, G., S. Muller, M. Beato, and G. Suske. 1992. Cloning by recognition site screening of two novel GT box binding proteins: a family of Sp1 related genes. Nucleic Acids Res **20**:5519-25.

78. Haltiwanger, R. S., M. A. Blomberg, and G. W. Hart. 1992. Glycosylation of nuclear and cytoplasmic proteins. Purification and characterization of a uridine diphospho-N-acetylglucosamine:polypeptide beta-N-acetylglucosaminyltransferase. J Biol Chem **267**:9005-13.

79. Haltiwanger, R. S., K. Grove, and G. A. Philipsberg. 1998. Modulation of O-linked N-acetylglucosamine levels on nuclear and cytoplasmic proteins in vivo using the peptide O-GlcNAc-beta-N-acetylglucosaminidase inhibitor O-(2-acetamido-2-deoxy-D-glucopyranosylidene)amino-N-phenylcarbamate. J Biol Chem **273**:3611-7.

80. Han, I., and J. E. Kudlow. 1997. Reduced O glycosylation of Sp1 is associated with increased proteasome susceptibility. Mol Cell Biol **17**:2550-8.

81. Han, Y., Y. B. Lin, W. An, J. Xu, H.-C. Yang, K. O'Connell, D. Dordai, J. D. Boeke, J. D. Siliciano, and R. F. Siliciano. 2008. Orientation-Dependent Regulation of Integrated HIV-1 Expression by Host Gene Transcriptional Readthrough. Cell Host & Microbe **4**:134-146.

82. Harada, F., G. G. Peters, and J. E. Dahlberg. 1979. The primer tRNA for Moloney murine leukemia virus DNA synthesis. Nucleotide sequence and aminoacylation of tRNAPro. J Biol Chem **254**:10979-85.

83. Harada, F., R. C. Sawyer, and J. E. Dahlberg. 1975. A primer ribonucleic acid for initiation of in vitro Rous sarcarcoma virus deoxyribonucleic acid synthesis. J Biol Chem **250**:3487-97.

84. Harrich, D., J. Garcia, F. Wu, R. Mitsuyasu, J. Gonazalez, and R. Gaynor. 1989. Role of SP1-binding domains in in vivo transcriptional regulation of the human immunodeficiency virus type 1 long terminal repeat. J Virol **63**:2585-91.

85. Hart, G. W. 1997. Dynamic O-linked glycosylation of nuclear and cytoskeletal proteins. Annual Review of Biochemistry **66**:315-335.

86. Hart, G. W., M. P. Housley, and C. Slawson. 2007. Cycling of O-linked [beta]-N-acetylglucosamine on nucleocytoplasmic proteins. Nature **446**:1017-1022.

87. Hart, G. W., L. K. Kreppel, F. I. Comer, C. S. Arnold, D. M. Snow, Z. Ye, X. Cheng, D. DellaManna, D. S. Caine, B. J. Earles, Y. Akimoto, R. N. Cole, and B. K. Hayes. 1996. O-GlcNAcylation of key nuclear and cytoskeletal proteins: reciprocity with O-phosphorylation and putative roles in protein multimerization. Glycobiology **6**:711-716.

References

88. **Havlir, D. V., M. C. Strain, M. Clerici, C. Ignacio, D. Trabattoni, P. Ferrante, and J. K. Wong.** 2003. Productive infection maintains a dynamic steady state of residual viremia in human immunodeficiency virus type 1-infected persons treated with suppressive antiretroviral therapy for five years. J Virol **77:**11212-9.

89. **Hazenberg, M. D., S. A. Otto, D. Hamann, M. T. Roos, H. Schuitemaker, R. J. de Boer, and F. Miedema.** 2003. Depletion of naive CD4 T cells by CXCR4-using HIV-1 variants occurs mainly through increased T-cell death and activation. AIDS **17:**1419-24.

90. **He, J., S. Choe, R. Walker, P. Di Marzio, D. O. Morgan, and N. R. Landau.** 1995. Human immunodeficiency virus type 1 viral protein R (Vpr) arrests cells in the G2 phase of the cell cycle by inhibiting p34cdc2 activity. J Virol **69:**6705-11.

91. **Heckel, D., N. Comtesse, N. Brass, N. Blin, K. D. Zang, and E. Meese.** 1998. Novel immunogenic antigen homologous to hyaluronidase in meningioma. Hum Mol Genet **7:**1859-72.

92. **Hermankova, M., S. C. Ray, C. Ruff, M. Powell-Davis, R. Ingersoll, R. T. D'Aquila, T. C. Quinn, J. D. Siliciano, R. F. Siliciano, and D. Persaud.** 2001. HIV-1 drug resistance profiles in children and adults with viral load of <50 copies/ml receiving combination therapy. JAMA **286:**196-207.

93. **Ho, H.-Y., M.-L. Cheng, S.-F. Weng, L. Chang, T.-T. Yeh, S.-R. Shih, and D. T.-Y. Chiu.** 2008. Glucose-6-phosphate dehydrogenase deficiency enhances enterovirus 71 infection. J Gen Virol **89:**2080-2089.

94. **Hoey, T., R. O. J. Weinzierl, G. Gill, J.-L. Chen, B. D. Dynlacht, and R. Tjian.** 1993. Molecular cloning and functional analysis of Drosophila TAF110 reveal properties expected of coactivators. Cell **72:**247-260.

95. **Holt, G. D., and G. W. Hart.** 1986. The subcellular distribution of terminal N-acetylglucosamine moieties. Localization of a novel protein-saccharide linkage, O-linked GlcNAc. J. Biol. Chem. **261:**8049-8057.

96. **Huang, J., F. Wang, E. Argyris, K. Chen, Z. Liang, H. Tian, W. Huang, K. Squires, G. Verlinghieri, and H. Zhang.** 2007. Cellular microRNAs contribute to HIV-1 latency in resting primary CD4+ T lymphocytes. Nat Med **13:**1241-7.

97. **Hung, J. J., Y. T. Wang, and W. C. Chang.** 2006. Sp1 deacetylation induced by phorbol ester recruits p300 to activate 12(S)-lipoxygenase gene transcription. Mol Cell Biol **26:**1770-85.

98. **Ikeda, M., and N. Kato.** 2007. Modulation of host metabolism as a target of new antivirals. Advanced Drug Delivery Reviews **59:**1277-1289.

99. **Inoki, K., T. Zhu, and K. L. Guan.** 2003. TSC2 mediates cellular energy response to control cell growth and survival. Cell **115:**577-90.

100. **Iyer, S. P., and G. W. Hart.** 2003. Dynamic nuclear and cytoplasmic glycosylation: enzymes of O-GlcNAc cycling. Biochemistry **42:**2493-9.

101. **Jackson, S. P., and R. Tjian.** 1988. O-glycosylation of eukaryotic transcription factors: implications for mechanisms of transcriptional regulation. Cell **55:**125-33.

102. **Jackson, S. P., and R. Tjian.** 1989. Purification and analysis of RNA polymerase II transcription factors by using wheat germ agglutinin affinity chromatography. Proc Natl Acad Sci U S A **86:**1781-5.

103. **James, L. R., D. Tang, A. Ingram, H. Ly, K. Thai, L. Cai, and J. W. Scholey.** 2002. Flux Through the Hexosamine Pathway Is a Determinant of Nuclear Factor {kappa}B- Dependent Promoter Activation. Diabetes **51:**1146-1156.

104. **Jeeninga, R. E., M. Hoogenkamp, M. Armand-Ugon, M. de Baar, K. Verhoef, and B. Berkhout.** 2000. Functional differences between the long terminal repeat transcriptional promoters of human immunodeficiency virus type 1 subtypes A through G. J Virol **74:**3740-51.

105. **Jiang, G., A. Espeseth, D. J. Hazuda, and D. M. Margolis.** 2007. c-Myc and Sp1 Contribute to Proviral Latency by Recruiting Histone Deacetylase 1 to the Human Immunodeficiency Virus Type 1 Promoter. J. Virol. **81:**10914-10923.

References

106. **Jinek, M., J. Rehwinkel, B. D. Lazarus, E. Izaurralde, J. A. Hanover, and E. Conti.** 2004. The superhelical TPR-repeat domain of O-linked GlcNAc transferase exhibits structural similarities to importin alpha. Nat Struct Mol Biol **11**:1001-7.

107. **Johnson, R. A., X. Wang, X. L. Ma, S. M. Huong, and E. S. Huang.** 2001. Human cytomegalovirus up-regulates the phosphatidylinositol 3-kinase (PI3-K) pathway: inhibition of PI3-K activity inhibits viral replication and virus-induced signaling. J Virol **75**:6022-32.

108. **Kadonaga, J. T., K. R. Carner, F. R. Masiarz, and R. Tjian.** 1987. Isolation of cDNA encoding transcription factor Sp1 and functional analysis of the DNA binding domain. Cell **51**:1079-1090.

109. **Kadonaga, J. T., A. J. Courey, J. Ladika, and R. Tjian.** 1988. Distinct regions of Sp1 modulate DNA binding and transcriptional activation. Science **242**:1566-70.

110. **Kaluza, G., C. Scholtissek, and R. Rott.** 1972. Inhibition of the Multiplication of Enveloped RNA-viruses by Glucosamine and 2-Deoxy-D-Glucose. J Gen Virol **14**:251-259.

111. **Kamine, J., and G. Chinnadurai.** 1992. Synergistic activation of the human immunodeficiency virus type 1 promoter by the viral Tat protein and cellular transcription factor Sp1. J Virol **66**:3932-6.

112. **Kamine, J., T. Subramanian, and G. Chinnadurai.** 1991. Sp1-dependent activation of a synthetic promoter by human immunodeficiency virus type 1 Tat protein. Proc Natl Acad Sci U S A **88**:8510-4.

113. **Kang, H. T., J. W. Ju, J. W. Cho, and E. S. Hwang.** 2003. Down-regulation of Sp1 activity through modulation of O-glycosylation by treatment with a low glucose mimetic, 2-deoxyglucose. J Biol Chem **278**:51223-31.

114. **Kang, S. M., A. C. Tran, M. Grilli, and M. J. Lenardo.** 1992. NF-kappa B subunit regulation in nontransformed CD4+ T lymphocytes. Science **256**:1452-6.

115. **Kao, S. Y., A. F. Calman, P. A. Luciw, and B. M. Peterlin.** 1987. Anti-termination of transcription within the long terminal repeat of HIV-1 by tat gene product. Nature **330**:489-93.

116. **Kearse, K. P., and G. W. Hart.** 1991. Lymphocyte activation induces rapid changes in nuclear and cytoplasmic glycoproteins. Proc Natl Acad Sci U S A **88**:1701-5.

117. **Keele, B. F., F. Van Heuverswyn, Y. Li, E. Bailes, J. Takehisa, M. L. Santiago, F. Bibollet-Ruche, Y. Chen, L. V. Wain, F. Liegeois, S. Loul, E. M. Ngole, Y. Bienvenue, E. Delaporte, J. F. Brookfield, P. M. Sharp, G. M. Shaw, M. Peeters, and B. H. Hahn.** 2006. Chimpanzee reservoirs of pandemic and nonpandemic HIV-1. Science **313**:523-6.

118. **Keembiyehetty, C. N., R. P. Candelaria, G. Majumdar, R. Raghow, A. Martinez-Hernandez, and S. S. Solomon.** 2002. Paradoxical regulation of Sp1 transcription factor by glucagon. Endocrinology **143**:1512-20.

119. **Kelly, W. G., M. E. Dahmus, and G. W. Hart.** 1993. RNA polymerase II is a glycoprotein. Modification of the COOH-terminal domain by O-GlcNAc. J. Biol. Chem. **268**:10416-10424.

120. **Keulen, W., M. Nijhuis, R. Schuurman, B. Berkhout, and C. Boucher.** 1997. Reverse transcriptase fidelity and HIV-1 variation. Science **275**:229; author reply 230-1.

121. **Kieffer, T. L., M. M. Finucane, R. E. Nettles, T. C. Quinn, K. W. Broman, S. C. Ray, D. Persaud, and R. F. Siliciano.** 2004. Genotypic analysis of HIV-1 drug resistance at the limit of detection: virus production without evolution in treated adults with undetectable HIV loads. J Infect Dis **189**:1452-65.

122. **Kihong, L., and C. Hyo-Ihl.** 2009. O-GlcNAc modification of Sp1 inhibits the functional interaction between Sp1 and Oct1. FEBS letters **583**:512-520.

123. **Kingsley, C., and A. Winoto.** 1992. Cloning of GT box-binding proteins: a novel Sp1 multigene family regulating T-cell receptor gene expression. Mol Cell Biol **12**:4251-61.

124. **Klenk, H. D., C. Scholtissek, and R. Rott.** 1972. Inhibition of glycoprotein biosynthesis of influenza virus by D-glucosamine and 2-deoxy-D-glucose. Virology **49**:723-34.

References

125. **Kollmar, R., K. A. Sukow, S. K. Sponagle, and P. J. Farnham.** 1994. Start site selection at the TATA-less carbamoyl-phosphate synthase (glutamine-hydrolyzing)/aspartate carbamoyltransferase/dihydroorotase promoter. J Biol Chem **269**:2252-7.

126. **Koster, J. C., M. S. Remedi, H. Qiu, C. G. Nichols, and P. W. Hruz.** 2003. HIV Protease Inhibitors Acutely Impair Glucose-Stimulated Insulin Release. Diabetes **52**:1695-1700.

127. **Koutsodontis, G., A. Moustakas, and D. Kardassis.** 2002. The role of Sp1 family members, the proximal GC-rich motifs, and the upstream enhancer region in the regulation of the human cell cycle inhibitor p21WAF-1/Cip1 gene promoter. Biochemistry **41**:12771-84.

128. **Kreppel, L. K., M. A. Blomberg, and G. W. Hart.** 1997. Dynamic glycosylation of nuclear and cytosolic proteins. Cloning and characterization of a unique O-GlcNAc transferase with multiple tetratricopeptide repeats. J Biol Chem **272**:9308-15.

129. **Kreppel, L. K., and G. W. Hart.** 1999. Regulation of a cytosolic and nuclear O-GlcNAc transferase. Role of the tetratricopeptide repeats. J Biol Chem **274**:32015-22.

130. **Kudlow, J. E.** 2006. Post-translational modification by O-GlcNAc: Another way to change protein function. Journal of Cellular Biochemistry **98**:1062-1075.

131. **Lacroix, I., C. Lipcey, J. Imbert, and B. Kahn-Perles.** 2002. Sp1 transcriptional activity is up-regulated by phosphatase 2A in dividing T lymphocytes. J Biol Chem **277**:9598-605.

132. **Laemmli, U. K.** 1970. Cleavage of structural proteins during the assembly of the head of bacteriophage T4. Nature **227**:680-5.

133. **Laspia, M. F., A. P. Rice, and M. B. Mathews.** 1989. HIV-1 Tat protein increases transcriptional initiation and stabilizes elongation. Cell **59**:283-92.

134. **Lassen, K., Y. Han, Y. Zhou, J. Siliciano, and R. F. Siliciano.** 2004. The multifactorial nature of HIV-1 latency. Trends in Molecular Medicine **10**:525-531.

135. **Laybourn, P. J., and M. E. Dahmus.** 1990. Phosphorylation of RNA polymerase IIA occurs subsequent to interaction with the promoter and before the initiation of transcription. J Biol Chem **265**:13165-73.

136. **Lazar, M. A.** 2003. Nuclear receptor corepressors. Nucl Recept Signal **1**:e001.

137. **Lazarus, B. D., D. C. Love, and J. A. Hanover.** 2006. Recombinant O-GlcNAc transferase isoforms: identification of O-GlcNAcase, yes tyrosine kinase, and tau as isoform-specific substrates. Glycobiology **16**:415-421.

138. **Lee, J.-A., D.-C. Suh, J.-E. Kang, M.-H. Kim, H. Park, M.-N. Lee, J.-M. Kim, H.-E. Roh, B.-N. Jeon, M.-Y. Yu, K.-Y. Choi, K. Y. Kim, and M.-W. Hur.** 2005. The transcriptional activity of Sp1 is regulated by molecular interactions between the zinc finger DNA binding domain and the inhibitory domain with corepressors, and this interaction is modulated by MEK. J. Biol. Chem.:M414134200.

139. **Lefebvre, T., C. Alonso, S. Mahboub, M. J. Dupire, J. P. Zanetta, M. L. Caillet-Boudin, and J. C. Michalski.** 1999. Effect of okadaic acid on O-linked N-acetylglucosamine levels in a neuroblastoma cell line. Biochim Biophys Acta **1472**:71-81.

140. **Legler, G., C. M. Muller-Platz, M. Mentges-Hettkamp, G. Pflieger, and E. Julich.** 1985. On the chemical basis of the Lowry protein determination. Anal Biochem **150**:278-87.

141. **Lehmann, M. H., J. Weber, O. Gastmann, and H. H. Sigusch.** 2002. Pseudogene-free amplification of human GAPDH cDNA. Biotechniques **33**:766, 769-70.

142. **Leonard, J., C. Parrott, A. J. Buckler-White, W. Turner, E. K. Ross, M. A. Martin, and A. B. Rabson.** 1989. The NF-kappa B binding sites in the human immunodeficiency virus type 1 long terminal repeat are not required for virus infectivity. J Virol **63**:4919-24.

143. **Letovsky, J., and W. S. Dynan.** 1989. Measurement of the binding of transcription factor Sp1 to a single GC

box recognition sequence. Nucleic Acids Res **17**:2639-53.

144. **Lewinski, M. K., D. Bisgrove, P. Shinn, H. Chen, C. Hoffmann, S. Hannenhalli, E. Verdin, C. C. Berry, J. R. Ecker, and F. D. Bushman.** 2005. Genome-wide analysis of chromosomal features repressing human immunodeficiency virus transcription. J Virol **79**:6610-9.

145. **Liao, M., Y. Zhang, and M. L. Dufau.** 2008. PKC{alpha}-induced Derepression of the Human LH Receptor Gene Transcription through ERK-mediated Release of HDAC1/Sin3A Corepressor Complex from Sp1 Sites. Mol Endocrinol:me.2008-0035.

146. **Love, D. C., and J. A. Hanover.** 2005. The Hexosamine Signaling Pathway: Deciphering the "O-GlcNAc Code". Sci. STKE **2005**:re13-.

147. **Lubas, W. A., D. W. Frank, M. Krause, and J. A. Hanover.** 1997. O-Linked GlcNAc Transferase Is a Conserved Nucleocytoplasmic Protein Containing Tetratricopeptide Repeats. J. Biol. Chem. **272**:9316-9324.

148. **Lubeseder-Martellato, C., E. Guenzi, A. Jorg, K. Topolt, E. Naschberger, E. Kremmer, C. Zietz, E. Tschachler, P. Hutzler, M. Schwemmle, K. Matzen, T. Grimm, B. Ensoli, and M. Stürzl.** 2002. Guanylate-binding protein-1 expression is selectively induced by inflammatory cytokines and is an activation marker of endothelial cells during inflammatory diseases. Am J Pathol **161**:1749-59.

149. **Lutz, N. W., N. Yahi, J. Fantini, and P. J. Cozzone.** 1997. Perturbations of glucose metabolism associated with HIV infection in human intestinal epithelial cells: a multinuclear magnetic resonance spectroscopy study. Aids **11**:147-55.

150. **Majello, B., P. De Luca, G. Hagen, G. Suske, and L. Lania.** 1994. Different members of the Sp1 multigene family exert opposite transcriptional regulation of the long terminal repeat of HIV-1. Nucleic Acids Res **22**:4914-21.

151. **Majumdar, G., A. Harmon, R. Candelaria, A. Martinez-Hernandez, R. Raghow, and S. S. Solomon.** 2003. O-glycosylation of Sp1 and transcriptional regulation of the calmodulin gene by insulin and glucagon. Am J Physiol Endocrinol Metab **285**:E584-91.

152. **Majumdar, G., A. Harrington, J. Hungerford, A. Martinez-Hernandez, I. C. Gerling, R. Raghow, and S. Solomon.** 2006. Insulin dynamically regulates calmodulin gene expression by sequential o-glycosylation and phosphorylation of sp1 and its subcellular compartmentalization in liver cells. J Biol Chem **281**:3642-50.

153. **Malim, M. H., J. Hauber, S. Y. Le, J. V. Maizel, and B. R. Cullen.** 1989. The HIV-1 rev trans-activator acts through a structured target sequence to activate nuclear export of unspliced viral mRNA. Nature **338**:254-7.

154. **Mariani, R., D. Chen, B. Schrofelbauer, F. Navarro, R. Konig, B. Bollman, C. Munk, H. Nymark-McMahon, and N. R. Landau.** 2003. Species-specific exclusion of APOBEC3G from HIV-1 virions by Vif. Cell **114**:21-31.

155. **Marin, M., A. Karis, P. Visser, F. Grosveld, and S. Philipsen.** 1997. Transcription factor Sp1 is essential for early embryonic development but dispensable for cell growth and differentiation. Cell **89**:619-28.

156. **Marshall, S., R. Okuyama, and J. M. Rumberger.** 2005. Turnover and characterization of UDP-N-acetylglucosaminyl transferase in a stably transfected HeLa cell line. Biochem Biophys Res Commun **332**:263-70.

157. **Mathe, G.** 1999. Human obesity and thinness, hyperlipidemia, hyperglycemia, and insulin resistance associated with HIV1 protease inhibitors. Prevention by alternating several antiproteases in short sequences. Biomed Pharmacother **53**:449-51.

158. **McAllister, J. J., D. Phillips, S. Millhouse, J. Conner, T. Hogan, H. L. Ross, and B. Wigdahl.** 2000. Analysis of the HIV-1 LTR NF-kappaB-proximal Sp site III: evidence for cell type-specific gene regulation and viral replication. Virology **274**:262-77.

159. **McClain, D. A., W. A. Lubas, R. C. Cooksey, M. Hazel, G. J. Parker, D. C. Love, and J. A. Hanover.** 2002. Altered glycan-dependent signaling induces insulin resistance and hyperleptinemia. PNAS **99**:10695-10699.

References

160. **McClain, D. A., A. J. Paterson, M. D. Roos, X. Wei, and J. E. Kudlow.** 1992. Glucose and glucosamine regulate growth factor gene expression in vascular smooth muscle cells. Proc Natl Acad Sci U S A **89:**8150-4.

161. **Medina, L., K. Grove, and R. S. Haltiwanger.** 1998. SV40 large T antigen is modified with O-linked N-acetylglucosamine but not with other forms of glycosylation. Glycobiology **8:**383-391.

162. **Mikovits, J. A., H. A. Young, P. Vertino, J. P. Issa, P. M. Pitha, S. Turcoski-Corrales, D. D. Taub, C. L. Petrow, S. B. Baylin, and F. W. Ruscetti.** 1998. Infection with human immunodeficiency virus type 1 upregulates DNA methyltransferase, resulting in de novo methylation of the gamma interferon (IFN-gamma) promoter and subsequent downregulation of IFN-gamma production. Mol Cell Biol **18:**5166-77.

163. **Mosmann, T.** 1983. Rapid colorimetric assay for cellular growth and survival: application to proliferation and cytotoxicity assays. J Immunol Methods **65:**55-63.

164. **Naitou, H., J. Mimaya, Y. Horikoshi, and T. Morita.** 1997. Quantitative detection of human immunodeficiency virus type 1 (HIV-1) RNA by PCR and use as a prognostic marker and for evaluating antiretroviral therapy. Biol Pharm Bull **20:**1317-20.

165. **Naschberger, E., T. Werner, A. B. Vicente, E. Guenzi, K. Topolt, R. Leubert, C. Lubeseder-Martellato, P. J. Nelson, and M. Stürzl.** 2004. Nuclear factor-kappaB motif and interferon-alpha-stimulated response element co-operate in the activation of guanylate-binding protein-1 expression by inflammatory cytokines in endothelial cells. Biochem J **379:**409-20.

166. **Nielsen, S. J., M. Praestegaard, H. F. Jorgensen, and B. F. Clark.** 1998. Different Sp1 family members differentially affect transcription from the human elongation factor 1 A-1 gene promoter. Biochem J **333** (Pt 3):511-7.

167. **Nishinaka, T., Y. H. Fu, L. I. Chen, K. Yokoyama, and R. Chiu.** 1997. A unique cathepsin-like protease isolated from CV-1 cells is involved in rapid degradation of retinoblastoma susceptibility gene product, RB, and transcription factor SP1. Biochim Biophys Acta **1351:**274-86.

168. **O'Donnell, N., N. E. Zachara, G. W. Hart, and J. D. Marth.** 2004. Ogt-Dependent X-Chromosome-Linked Protein Glycosylation Is a Requisite Modification in Somatic Cell Function and Embryo Viability. Mol. Cell. Biol. **24:**1680-1690.

169. **Ortiz, B. D., A. M. Krensky, and P. J. Nelson.** 1996. Kinetics of transcription factors regulating the RANTES chemokine gene reveal a developmental switch in nuclear events during T-lymphocyte maturation. Mol Cell Biol **16:**202-10.

170. **Pantaleo, G., C. Graziosi, and A. S. Fauci.** 1993. The Immunopathogenesis of Human Immunodeficiency Virus Infection. N Engl J Med **328:**327-335.

171. **Parent, M., T. M. C. Yung, A. Rancourt, E. L. Y. Ho, S. Vispe, F. Suzuki-Matsuda, A. Uehara, T. Wada, H. Handa, and M. S. Satoh.** 2005. Poly(ADP-ribose) Polymerase-1 Is a Negative Regulator of HIV-1 Transcription through Competitive Binding to TAR RNA with Tat{middle dot}Positive Transcription Elongation Factor b (p-TEFb) Complex. J. Biol. Chem. **280:**448-457.

172. **Park, S. Y., J. Ryu, and W. Lee.** 2005. O-GlcNAc modification on IRS-1 and Akt2 by PUGNAc inhibits their phosphorylation and induces insulin resistance in rat primary adipocytes. Exp Mol Med **37:**220-9.

173. **Parrott, C., T. Seidner, E. Duh, J. Leonard, T. S. Theodore, A. Buckler-White, M. A. Martin, and A. B. Rabson.** 1991. Variable role of the long terminal repeat Sp1-binding sites in human immunodeficiency virus replication in T lymphocytes. J Virol **65:**1414-9.

174. **Pascal, E., and R. Tjian.** 1991. Different activation domains of Sp1 govern formation of multimers and mediate transcriptional synergism. Genes Dev **5:**1646-56.

175. **Patterson, S., and S. C. Knight.** 1987. Susceptibility of human peripheral blood dendritic cells to infection by human immunodeficiency virus. J Gen Virol **68** (Pt 4):1177-81.

176. **Patti, M. E., A. Virkamaki, E. J. Landaker, C. R. Kahn, and H. Yki-Jarvinen.** 1999. Activation of the hexosamine pathway by glucosamine in vivo induces insulin resistance of early postreceptor insulin signaling

events in skeletal muscle. Diabetes **48**:1562-71.

177. **Pereira, L. A., K. Bentley, A. Peeters, M. J. Churchill, and N. J. Deacon.** 2000. A compilation of cellular transcription factor interactions with the HIV-1 LTR promoter. Nucleic Acids Res **28**:663-8.

178. **Perelson, A. S., P. Essunger, Y. Cao, M. Vesanen, A. Hurley, K. Saksela, M. Markowitz, and D. D. Ho.** 1997. Decay characteristics of HIV-1-infected compartments during combination therapy. Nature **387**:188-91.

179. **Perkins, N. D., N. L. Edwards, C. S. Duckett, A. B. Agranoff, R. M. Schmid, and G. J. Nabel.** 1993. A cooperative interaction between NF-kappa B and Sp1 is required for HIV-1 enhancer activation. Embo J **12**:3551-8.

180. **Persaud, D., G. K. Siberry, A. Ahonkhai, J. Kajdas, D. Monie, N. Hutton, D. C. Watson, T. C. Quinn, S. C. Ray, and R. F. Siliciano.** 2004. Continued production of drug-sensitive human immunodeficiency virus type 1 in children on combination antiretroviral therapy who have undetectable viral loads. J Virol **78**:968-79.

181. **Persaud, D., Y. Zhou, J. M. Siliciano, and R. F. Siliciano.** 2003. Latency in human immunodeficiency virus type 1 infection: no easy answers. J Virol **77**:1659-65.

182. **Popov, S., M. Rexach, G. Zybarth, N. Reiling, M. A. Lee, L. Ratner, C. M. Lane, M. S. Moore, G. Blobel, and M. Bukrinsky.** 1998. Viral protein R regulates nuclear import of the HIV-1 pre-integration complex. EMBO J **17**:909-17.

183. **Pujalte, J. M., E. P. Llavore, and F. R. Ylescupidez.** 1980. Double-blind clinical evaluation of oral glucosamine sulphate in the basic treatment of osteoarthrosis. Curr Med Res Opin **7**:110-14.

184. **Quivy, V., E. Adam, Y. Collette, D. Demonte, A. Chariot, C. Vanhulle, B. Berkhout, R. Castellano, Y. de Launoit, A. Burny, J. Piette, V. Bours, and C. Van Lint.** 2002. Synergistic activation of human immunodeficiency virus type 1 promoter activity by NF-kappaB and inhibitors of deacetylases: potential perspectives for the development of therapeutic strategies. J Virol **76**:11091-103.

185. **Rao, J., F. Zhang, R. J. Donnelly, N. L. Spector, and G. P. Studzinski.** 1998. Truncation of Sp1 transcription factor by myeloblastin in undifferentiated HL60 cells. J Cell Physiol **175**:121-8.

186. **Ratner, L., W. Haseltine, R. Patarca, K. J. Livak, B. Starcich, S. F. Josephs, E. R. Doran, J. A. Rafalski, E. A. Whitehorn, K. Baumeister, and et al.** 1985. Complete nucleotide sequence of the AIDS virus, HTLV-III. Nature **313**:277-84.

187. **Reeves, J. D., and R. W. Doms.** 2002. Human immunodeficiency virus type 2. J Gen Virol **83**:1253-65.

188. **Reginster, J. Y., O. Bruyere, G. Fraikin, and Y. Henrotin.** 2005. Current concepts in the therapeutic management of osteoarthritis with glucosamine. Bull Hosp Jt Dis **63**:31-6.

189. **Rickers, A., N. Peters, V. Badock, R. Beyaert, P. Vandenabeele, B. Dorken, and K. Bommert.** 1999. Cleavage of transcription factor SP1 by caspases during anti-IgM-induced B-cell apoptosis. Eur J Biochem **261**:269-74.

190. **Ringler, D. J., M. S. Wyand, D. G. Walsh, J. J. MacKey, L. V. Chalifoux, M. Popovic, A. A. Minassian, P. K. Sehgal, M. D. Daniel, R. C. Desrosiers, and et al.** 1989. Cellular localization of simian immunodeficiency virus in lymphoid tissues. I. Immunohistochemistry and electron microscopy. Am J Pathol **134**:373-83.

191. **Roos, M. D., K. Su, J. R. Baker, and J. E. Kudlow.** 1997. O glycosylation of an Sp1-derived peptide blocks known Sp1 protein interactions. Mol Cell Biol **17**:6472-80.

192. **Roos, M. D., W. Xie, K. Su, J. A. Clark, X. Yang, E. Chin, A. J. Paterson, and J. E. Kudlow.** 1998. Streptozotocin, an analog of N-acetylglucosamine, blocks the removal of O-GlcNAc from intracellular proteins. Proc Assoc Am Physicians **110**:422-32.

193. **Rosen, C. A., J. G. Sodroski, and W. A. Haseltine.** 1985. The location of cis-acting regulatory sequences in the human T cell lymphotropic virus type III (HTLV-III/LAV) long terminal repeat. Cell **41**:813-823.

194. **Saffer, J. D., S. P. Jackson, and M. B. Annarella.** 1991. Developmental expression of Sp1 in the mouse. Mol

Cell Biol **11:**2189-99.

195. **Salter, R. D., D. N. Howell, and P. Cresswell.** 1985. Genes regulating HLA class I antigen expression in T-B lymphoblast hybrids. Immunogenetics **21:**235-46.

196. **Sander, G., A. Konrad, M. Thurau, E. Wies, R. Leubert, E. Kremmer, H. Dinkel, T. Schulz, F. Neipel, and M. Stürzl.** 2008. Intracellular localization map of human herpesvirus 8 proteins. J Virol **82:**1908-22.

197. **Sanger, F., S. Nicklen, and A. R. Coulson.** 1977. DNA sequencing with chain-terminating inhibitors. Proc Natl Acad Sci U S A **74:**5463-7.

198. **Schaft, N., J. Dorrie, I. Muller, V. Beck, S. Baumann, T. Schunder, E. Kampgen, and G. Schuler.** 2006. A new way to generate cytolytic tumor-specific T cells: electroporation of RNA coding for a T cell receptor into T lymphocytes. Cancer Immunol Immunother **55:**1132-41.

199. **Schaft, N., J. Dorrie, P. Thumann, V. E. Beck, I. Muller, E. S. Schultz, E. Kampgen, D. Dieckmann, and G. Schuler.** 2005. Generation of an optimized polyvalent monocyte-derived dendritic cell vaccine by transfecting defined RNAs after rather than before maturation. J Immunol **174:**3087-97.

200. **Schneider, U., H. U. Schwenk, and G. Bornkamm.** 1977. Characterization of EBV-genome negative "null" and "T" cell lines derived from children with acute lymphoblastic leukemia and leukemic transformed non-Hodgkin lymphoma. Int J Cancer **19:**621-6.

201. **Scholtissek, C., R. Rott, G. Hau, and G. Kaluza.** 1974. Inhibition of the multiplication of vesicular stomatitis and Newcastle disease virus by 2-deoxy-d-glucose. J Virol **13:**1186-93.

202. **Scholtissek, C., R. Rott, and H. D. Klenk.** 1975. Two different mechanisms of the inhibition of the multiplication of enveloped viruses by glucosamine. Virology **63:**191-200.

203. **Schubert, U., A. V. Ferrer-Montiel, M. Oblatt-Montal, P. Henklein, K. Strebel, and M. Montal.** 1996. Identification of an ion channel activity of the Vpu transmembrane domain and its involvement in the regulation of virus release from HIV-1-infected cells. FEBS Lett **398:**12-8.

204. **Schwartz, M. D., R. J. Geraghty, and A. T. Panganiban.** 1996. HIV-1 particle release mediated by Vpu is distinct from that mediated by p6. Virology **224:**302-9.

205. **Shafi, R., S. P. Iyer, L. G. Ellies, N. O'Donnell, K. W. Marek, D. Chui, G. W. Hart, and J. D. Marth.** 2000. The O-GlcNAc transferase gene resides on the X chromosome and is essential for embryonic stem cell viability and mouse ontogeny. Proc Natl Acad Sci U S A **97:**5735-9.

206. **Shen, L., and R. F. Siliciano.** 2008. Viral reservoirs, residual viremia, and the potential of highly active antiretroviral therapy to eradicate HIV infection. Journal of Allergy and Clinical Immunology **122:**22-28.

207. **Shlomai, A., and Y. Shaul.** 2008. The "metabolovirus" model of hepatitis B virus suggests nutritional therapy as an effective anti-viral weapon. Medical Hypotheses **71:**53-57.

208. **Simon, V., and D. D. Ho.** 2003. HIV-1 dynamics in vivo: implications for therapy. Nat Rev Microbiol **1:**181-90.

209. **Slawson, C., and G. W. Hart.** 2003. Dynamic interplay between O-GlcNAc and O-phosphate: the sweet side of protein regulation. Current Opinion in Structural Biology **13:**631-636.

210. **Slawson, C., M. P. Housley, and G. W. Hart.** 2006. O-GlcNAc cycling: How a single sugar post-translational modification is changing the Way We think about signaling networks. Journal of Cellular Biochemistry **97:**71-83.

211. **Spengler, M. L., and M. G. Brattain.** 2006. Sumoylation inhibits cleavage of Sp1 N-terminal negative regulatory domain and inhibits Sp1-dependent transcription. J Biol Chem **281:**5567-74.

212. **Spengler, M. L., L. W. Guo, and M. G. Brattain.** 2008. Phosphorylation mediates Sp1 coupled activities of proteolytic processing, desumoylation and degradation. Cell Cycle **7:**623-30.

References

213. **Sprung, R., A. Nandi, Y. Chen, S. C. Kim, D. Barma, J. R. Falck, and Y. Zhao.** 2005. Tagging-via-substrate strategy for probing O-GlcNAc modified proteins. J Proteome Res **4**:950-7.

214. **Starcich, B., L. Ratner, S. F. Josephs, T. Okamoto, R. C. Gallo, and F. Wong-Staal.** 1985. Characterization of long terminal repeat sequences of HTLV-III. Science **227**:538-40.

215. **Stevenson, M.** 2003. HIV-1 pathogenesis. Nat Med **9**:853-60.

216. **Stevenson, M., S. Haggerty, C. A. Lamonica, C. M. Meier, S. K. Welch, and A. J. Wasiak.** 1990. Integration is not necessary for expression of human immunodeficiency virus type 1 protein products. J Virol **64**:2421-5.

217. **Stevenson, M., T. L. Stanwick, M. P. Dempsey, and C. A. Lamonica.** 1990. HIV-1 replication is controlled at the level of T cell activation and proviral integration. EMBO J **9**:1551-60.

218. **Stumptner-Cuvelette, P., S. Morchoisne, M. Dugast, S. Le Gall, G. Raposo, O. Schwartz, and P. Benaroch.** 2001. HIV-1 Nef impairs MHC class II antigen presentation and surface expression. Proc Natl Acad Sci U S A **98**:12144-9.

219. **Su, K., M. D. Roos, X. Yang, I. Han, A. J. Paterson, and J. E. Kudlow.** 1999. An N-terminal region of Sp1 targets its proteasome-dependent degradation in vitro. J Biol Chem **274**:15194-202.

220. **Suné, C., and M. A. Garcia-Blanco.** 1995. Sp1 transcription factor is required for in vitro basal and Tat-activated transcription from the human immunodeficiency virus type 1 long terminal repeat. J Virol **69**:6572-6.

221. **Suske, G.** 1999. The Sp-family of transcription factors. Gene **238**:291-300.

222. **Suzuki, T., A. Kimura, R. Nagai, and M. Horikoshi.** 2000. Regulation of interaction of the acetyltransferase region of p300 and the DNA-binding domain of Sp1 on and through DNA binding. Genes Cells **5**:29-41.

223. **Tenner-Racz, K., P. Racz, H. Schmidt, M. Dietrich, P. Kern, A. Louie, S. Gartner, and M. Popovic.** 1988. Immunohistochemical, electron microscopic and in situ hybridization evidence for the involvement of lymphatics in the spread of HIV-1. AIDS **2**:299-309.

224. **Thurau, M., G. Marquardt, N. Gonin-Laurent, K. Weinlander, E. Naschberger, R. Jochmann, K. R. Alkharsah, T. F. Schulz, M. Thome, F. Neipel, and M. Stürzl.** 2008. The Viral Inhibitor of Apoptosis vFLIP/K13 Protects Endothelial Cells against Superoxide-induced Cell Death. J Virol **83**:598-611.

225. **Toleman, C., A. J. Paterson, T. R. Whisenhunt, and J. E. Kudlow.** 2004. Characterization of the histone acetyltransferase (HAT) domain of a bifunctional protein with activable O-GlcNAcase and HAT activities. J Biol Chem **279**:53665-73.

226. **Tomar, R. H.** 1994. Breaking the asymptomatic phase of HIV-1 infection. J Clin Lab Anal **8**:116-9.

227. **Torres, C. R., and G. W. Hart.** 1984. Topography and polypeptide distribution of terminal N-acetylglucosamine residues on the surfaces of intact lymphocytes. Evidence for O-linked GlcNAc. J Biol Chem **259**:3308-17.

228. **Trono, D.** 1995. HIV accessory proteins: Leading roles for the supporting cast. Cell **82**:189-192.

229. **Uetsuki, T., A. Naito, S. Nagata, and Y. Kaziro.** 1989. Isolation and characterization of the human chromosomal gene for polypeptide chain elongation factor-1 alpha. J Biol Chem **264**:5791-8.

230. **UNAIDS.** 2008. Report on the global HIV/AIDS epidemic 2008: executive summary. UNAIDS/08.27E / JC1511E.

231. **Vanderford, N. L., S. S. Andrali, and S. Özcan.** 2006. Glucose induces MafA expression in pancreatic beta cell lines via the hexosamine biosynthetic pathway. J. Biol. Chem.:M605064200.

232. **Vettese-Dadey, M., P. A. Grant, T. R. Hebbes, C. Crane- Robinson, C. D. Allis, and J. L. Workman.** 1996. Acetylation of histone H4 plays a primary role in enhancing transcription factor binding to nucleosomal DNA in vitro. EMBO J **15**:2508-18.

References

233. **Wakabayashi-Ito, N., and S. Nagata.** 1994. Characterization of the regulatory elements in the promoter of the human elongation factor-1 alpha gene. J Biol Chem **269**:29831-7.

234. **Wang, Y.-T., J.-Y. Chuang, M.-R. Shen, W.-B. Yang, W.-C. Chang, and J.-J. Hung.** 2008. Sumoylation of Specificity Protein 1 Augments Its Degradation by Changing the Localization and Increasing the Specificity Protein 1 Proteolytic Process. Journal of Molecular Biology **380**:869-885.

235. **Weinberg, J. B., T. J. Matthews, B. R. Cullen, and M. H. Malim.** 1991. Productive human immunodeficiency virus type 1 (HIV-1) infection of nonproliferating human monocytes. J Exp Med **174**:1477-82.

236. **Weinländer, K., E. Naschberger, M. H. Lehmann, P. Tripal, W. Paster, H. Stockinger, C. Hohenadl, and M. Stürzl.** 2008. Guanylate binding protein-1 inhibits spreading and migration of endothelial cells through induction of integrin alpha4 expression. FASEB J **22**:4168-78.

237. **Wells, L., Y. Gao, J. A. Mahoney, K. Vosseller, C. Chen, A. Rosen, and G. W. Hart.** 2002. Dynamic O-glycosylation of nuclear and cytosolic proteins: further characterization of the nucleocytoplasmic beta-N-acetylglucosaminidase, O-GlcNAcase. J Biol Chem **277**:1755-61.

238. **Wells, L., and G. W. Hart.** 2003. O-GlcNAc turns twenty: functional implications for post-translational modification of nuclear and cytosolic proteins with a sugar. FEBS Letters **546**:154-158.

239. **Whelan, S. A., M. D. Lane, and G. W. Hart.** 2008. Regulation of the O-Linked {beta}-N-Acetylglucosamine Transferase by Insulin Signaling. J. Biol. Chem. **283**:21411-21417.

240. **Whisenhunt, T. R., X. Yang, D. B. Bowe, A. J. Paterson, B. A. Van Tine, and J. E. Kudlow.** 2006. Disrupting the enzyme complex regulating O-GlcNAcylation blocks signaling and development. Glycobiology **16**:551-63.

241. **Wierstra, I.** 2008. Sp1: Emerging roles--Beyond constitutive activation of TATA-less housekeeping genes. Biochemical and Biophysical Research Communications **372**:1-13.

242. **Williams, S. A., L. F. Chen, H. Kwon, C. M. Ruiz-Jarabo, E. Verdin, and W. C. Greene.** 2006. NF-kappaB p50 promotes HIV latency through HDAC recruitment and repression of transcriptional initiation. Embo J **25**:139-49.

243. **Wolfe, S. A., L. Nekludova, and C. O. Pabo.** 2000. DNA Recognition By Cys2His2 Zinc Finger Proteins. Annual Review of Biophysics and Biomolecular Structure **29**:183.

244. **Wolfe, S. A., L. Nekludova, and C. O. Pabo.** 2000. DNA recognition by Cys2His2 zinc finger proteins. Annu Rev Biophys Biomol Struct **29**:183-212.

245. **Wong, J. K., M. Hezareh, H. F. Gunthard, D. V. Havlir, C. C. Ignacio, C. A. Spina, and D. D. Richman.** 1997. Recovery of replication-competent HIV despite prolonged suppression of plasma viremia. Science **278**:1291-5.

246. **Wu, Y.** 2004. HIV-1 gene expression: lessons from provirus and non-integrated DNA. Retrovirology **1**:13.

247. **Yang, X., K. Su, M. D. Roos, Q. Chang, A. J. Paterson, and J. E. Kudlow.** 2001. O-linkage of N-acetylglucosamine to Sp1 activation domain inhibits its transcriptional capability. Proc Natl Acad Sci U S A **98**:6611-6.

248. **Yang, X., F. Zhang, and J. E. Kudlow.** 2002. Recruitment of O-GlcNAc transferase to promoters by corepressor mSin3A: coupling protein O-GlcNAcylation to transcriptional repression. Cell **110**:69-80.

249. **Yedavalli, V. S., M. Benkirane, and K. T. Jeang.** 2003. Tat and trans-activation-responsive (TAR) RNA-independent induction of HIV-1 long terminal repeat by human and murine cyclin T1 requires Sp1. J Biol Chem **278**:6404-10.

250. **Yeni, P.** 2006. Update on HAART in HIV. J Hepatol **44**:S100-3.

251. **Ylisastigui, L., J. J. Coull, V. C. Rucker, C. Melander, R. J. Bosch, S. J. Brodie, L. Corey, D. L. Sodora,**

References

P. B. Dervan, and D. M. Margolis. 2004. Polyamides reveal a role for repression in latency within resting T cells of HIV-infected donors. J Infect Dis **190**:1429-37.

252. **Zachara, N. E., and G. W. Hart.** 2002. The emerging significance of O-GlcNAc in cellular regulation. Chem Rev **102**:431-8.

253. **Zachara, N. E., and G. W. Hart.** 2004. O-GlcNAc a sensor of cellular state: the role of nucleocytoplasmic glycosylation in modulating cellular function in response to nutrition and stress. Biochimica et Biophysica Acta (BBA) - General Subjects **1673**:13-28.

254. **Zaniolo, K., S. Desnoyers, S. Leclerc, and S. L. Guerin.** 2007. Regulation of poly(ADP-ribose) polymerase-1 (PARP-1) gene expression through the post-translational modification of Sp1: a nuclear target protein of PARP-1. BMC Mol Biol **8**:96.

255. **Zaniolo, K., A. Rufiange, S. Leclerc, S. Desnoyers, and S. L. Guerin.** 2005. Regulation of the poly(ADP-ribose) polymerase-1 gene expression by the transcription factors Sp1 and Sp3 is under the influence of cell density in primary cultured cells. Biochem J **389**:423-33.

256. **Zannetti, C., M. Mondini, M. De Andrea, P. Caposio, E. Hara, G. Peters, G. Gribaudo, M. Gariglio, and S. Landolfo.** 2006. The expression of p16INK4a tumor suppressor is upregulated by human cytomegalovirus infection and required for optimal viral replication. Virology **349**:79-86.

257. **Zhang, F., K. Su, X. Yang, D. B. Bowe, A. J. Paterson, and J. E. Kudlow.** 2003. O-GlcNAc modification is an endogenous inhibitor of the proteasome. Cell **115**:715-25.

258. **Zhang, Y., and M. L. Dufau.** 2002. Silencing of Transcription of the Human Luteinizing Hormone Receptor Gene by Histone Deacetylase-mSin3A Complex. J. Biol. Chem. **277**:33431-33438.

259. **Zheng, X. L., S. Matsubara, C. Diao, M. D. Hollenberg, and N. C. Wong.** 2001. Epidermal growth factor induction of apolipoprotein A-I is mediated by the Ras-MAP kinase cascade and Sp1. J Biol Chem **276**:13822-9.

7. Abbreviations

2-DG	2-deoxy glucose
β-ME	β-mercaptoethanol
µg	microgram(s)
µl	microliter(s)
AIDS	Acquired Immune Deficiency Syndrome
Amp	ampicillin
AP1	activator protein 1
APS	ammonium persulfate
ATP	adenosine triphosphate
bp	base pairs
BSA	bovine serum albumin
BSB	band shift buffer
$CaCl_2$	calcium chloride
cDNA	complementary DNA
CMV	cytomegalovirus
cpm	counts per minute
ddH_2O	double distilled water
dEGFP	destabilized enhanced green fluorescent protein
DEPC	diethylene pyrocarbonate
DMEM	Dulbecco`s Modified Eagle Medium
DMSO	dimethyl sulfoxide
DNA	desoxyribonucleic acid
dNTP	desoxyribonucleotide triphosphate
dsDNA	double stranded DNA
DTT	dithiotreitol
ECL	enhanced chemiluminescence
EDTA	ethylenediamine tetraacetic acid
EF1α	elongation factor 1α
EGFP	enhanced green fluorescent protein
FACS	fluorescence activated cell sorting
FCS	fetal calf sera
Fig	Figure
fwd	forward
g	gram(ss) or gravitation constant
GAPDH	Glycerinaldehyd-3-Phosphat-dehydrogenase
GFAT	glutamine:fructose-6-phosphate amidotransferase
GFP	green fluorescent protein
GlcN	glucosamine
h	hour(s)
HAART	highly active antiretroviral therapy
HAT	histone acetyltransferase
HBP	hexosamine biosynthesis pathway
HBS	HEPES buffered saline
HDAC	histone deacetylase
HEK	human embryonic kidney
HeLa	cervix carcinoma cell line isolated from *Henrietta Lacks*
HEPES	4-(2-hydroxyethyl)-1-piperazineethanesulfonic acid
HIV-1, -2	human immunodeficiency virus type-1, type-2

Abbreviations

HRP	horseradish peroxidase
IgG	immunoglobulin G
IL	interleukin
IP	immunoprecipitation
IR	insulin receptor
IRS	insulin receptor substrate
kana	kanamycin
kb	kilo base pairs
kDa	kilodalton
L	liter(s)
LB	lysogeny broth
LDH	lactate dehydrogenase
LTR	long terminal repeat
Luc	luciferase
mA	milliampere
MCS	multiple cloning site
mg	milligram(s)
$MgCl_2$	magnesium chloride
$MgSO_4$	magnesium sulfate
min	minute(s)
ml	milliliter(s)
mM	millimolar
mOGT	mitochondrial OGT
MOPS	4-Morpholinepropanesulfonic acid
ms	millisecond(s)
mut	mutated
Na_2HPO_4	sodium hydrogen phosphate
NaCl	sodium chloride
NaN_3	sodium azide
NCOAT	nuclear and cytoplasmic *O*-GlcNAcase and histone acetyltransferase
ncOGT	nucleocytoplasmic OGT
NFAT	nuclear factor of activated T cells
NF-κB	nuclear factor-kappa B
ng	nanogram(s)
nm	nanometer(s)
nt	nucleotide
OD_x	optical density, at wavelength λ = x nm
O-GlcNAc	*O*-liked N-Acetyl-D-Glucosamine
O-GlcNAcase	*O*-GlcNAc hexosaminidase
OGT	*O*-GlcNAc transferase
P-value	probability
PAGE	polyacrylamide gel electrophoresis
PBS	phosphate buffered saline
PCR	polymerase chain reaction
pH	power of hydrogen
PNK	polynucleotide kinase
P-TEFb	positive transcription elongation factor b
PVDF	polyvinylidene fluoride
Q	glutamine

Abbreviations

res	rescue
rev	reverse
RNA	ribonucleic acid
rpm	revolutions per minute
RPMI	Roswell Park Memorial Institute
RT	room temperature / reverse transcriptase
S / Ser	serine
SDS	sodium dodecyl sulfate
sec	second(s)
siRNA	short interfering RNA
SOC	super optimal broth with catabolite repression
sOGT	small isoform of OGT
Sp1	specificity protein 1
ssRNA	single stranded RNA
STE	sodium-tris-EDTA
T / Thr	threonine
Tab.	table
TAE	tris-acetate-EDTA
TAF	TBP associated factor
TAR	transactivation response element
Tat	transactivator of transcription
TBP	TATA-binding protein
TBE	tris-boric acid-EDTA
TBq	terrabecquerel
TEMED	N, N, N´, N´-tetramethylethylenediamine
TF	transcription factor
TFIID	transcription factor II D
TPR	tetratricopeptide repeat
Tris	tris(hydroxymethyl)-aminomethane
U	unit(s)
UDP-GlcNAc	uridine diphosphate N-acetyl D-glucosamine
UV	ultraviolet
V	Volt
VLE	very low endotoxin
VSV	vesicular stomatitis virus
w/v	weight per volume
wt	wild-type

8. Publications

PARTS OF THIS WORK WERE PUBLISHED IN:

Jochmann R, Thurau M, Jung S, Hofmann C, Naschberger E, Kremmer E, Harrer T, Miller M, Schaft N, Stürzl M (2009). *O*-linked N-Acetylglucosaminylation of Sp1 Inhibits the Human Immunodeficiency Virus Type-1 Promoter. J Virol, Vol 83(8), in press.

OTHER PUBLICATIONS:

Konrad A, Wies E, Thurau M, Marquardt G, Naschberger E, Hentschel S, **Jochmann R**, Schulz TF, Erfle H, Brors B, Lausen B, Neipel F, Stürzl M (2009). A systems biology approach to identify combination effects of HHV-8 genes on NF-{kappa}B activation. J Virol, Vol 83(6).

Thurau M, Marquardt G, Gonin-Laurent N, Weinländer K, Naschberger E, **Jochmann R**, Alkharsah KR, Schulz TF, Thome M, Neipel F, Stürzl M (2009). Viral inhibitor of apoptosis vFLIP/K13 protects endothelial cells against superoxide-induced cell death. J Virol, Vol 83(2).

Schregel V, Auerochs S, **Jochmann R**, Maurer K, Stamminger T, Marschall M (2007). Mapping of a self-interaction domain of the cytomegalovirus protein kinase pUL97. J Gen Virol, Vol 88(Pt 2).

Marschall M, Marzi A, aus dem Siepen P, **Jochmann R**, Kalmer M, Auerochs S, Lischka P, Leis M, Stamminger T (2005). Cellular p32 recruits cytomegalovirus kinase pUL97 to redistribute the nuclear lamina. J Biol Chem, Vol 280(39).

9. Acknowledgements

I want to thank all the people, whose support was essential for the progression of this work:

Prof. Dr. Michael Stürzl *for accepting me in his laboratory, for the consistent support during this work, and for giving me fresh impetus when I mostly needed it. He further gave me the opportunity to broaden my mind by encouraging me to participate at the Meeting of Nobel Prize Winners in Chemistry at Lindau and by supporting my interest in other topics apart from HIV-1.*

Prof. Dr. Thomas Winkler *for accepting the tutorial of this work.*

Prof. Dr. Werner Hohenberger *for the excellent working conditions provided in the lab.*

Dr. Mathias Thurau *for his continuous support during my Ph.D., and for his appreciative words for a good experimental idea or for a well written paragraph.*

Prof. Bernhard Fleckenstein, PD Dr. Brigitte Biesinger *and my colleagues from the* **Graduiertenkolleg 1071**. *I appreciate the huge experience I gained during the retreats, seminars and workshops, which would not have taken place without them.*

Manuel López-Cabrera *(University of Molecular Biology, University Hospital of Princesa, Madrid, Spain) for kindly providing us the pXP1-LTR-plasmids.* **Susan Jung** *(Institute of Clinical and Molecular Virology) for providing the pNL4-3, pNL4-3$_{LucR-E-}$ plasmids, and for her patience in answering all my questions concerning HIV-1 infection protocols.* **Christian Hofmann, Dr. Niels Schaft and Dr. Jan Dorrie** *(all from the Department of Dermatology) for the isolation of $CD4^+$ primary T cells and for the support in transfecting these cells with OGT mRNA.*

Dr. Elisabeth Kremmer *(Institute of Molecular Immunology, Helmholtz Center, Munich, Germany) for the production of the monoclonal antibody against HIV-1 Tat.* **Prof. Dr. Christoph Garlichs** *(Department of Medicine II, Cardiology and Angiology) for the opportunity to use the FACSCalibur.* **Prof. Dr. Ulrich Schubert** *(Institute of Clinical and Molecular Virology) for providing us the Gene Pulser XCell.*

Dr. Elisabeth Naschberger *for her kind support and good experimental ideas, and for introducing me in the EMSA method.*

Prof. Dr. Thomas Stamminger *for mentoring this work.* **Prof. Dr. Manfred Marschall** *and* **Dr. Peter Lischka** *for teaching me curiosity and the fascination of scientific research.*

Thank to the blood donors.

However, a Ph.D. student would feel pretty lost without colleagues. Therefore I'm happy to have had **Andreas Konrad, Elisabeth Kuhn** *and* **Dr. Kristina Weinländer** *around me, with whom I shared the office. Thank you for your support and for lending me your ears in every situation and for every kind of problem.*

A special thank to **Helene Demund, Nicole Fischer, Matthias Hammon, Christina von Kleinsorgen, Melanie Nurtsch, Bernd Öchsner, Alexander Schäfer, Stefanie Scholz,** *and* **Dr. Philipp Tripal** *for the funny moments in the lab and to* **Dr. Michael Bauer** *and* **Gaby Marquardt** *for sharing their lab experience. I further thank* **Dr. Nathalie Gonin-Laurent** *for the opportunity to refresh my French and for sharing her knowledge concerning the structure of the research community in France.*

I further want to thank **Susanne Reed** *for organizing the office, for her help concerning bureaucracy and especially for the good vibrations she spread around. A special thank to* **Mahimaidos Manoharan** *for solving every accruing special problem in the lab and for introducing me, together with* **Priya Chudasama**, *in the culture of India. I thank* **Gertrud Hoffmann** *for reminding me that there are also other problems apart from transfection efficiencies and cloning, and* **Tamara Weber** *for organizing the scullery and for her continuous smiling that welcomed me every morning.*

A heartily thank to **Manuel Weikert** *and to my earlier teacher* **Dr. Hans Krautblatter**, *who both taught me the fascination of life and evolution, and evoked my interest in biology.*

I'm lucky to have met **Mirjam Metzner**, *who was not only a colleague but also a great friend. With her I shared the ups and the downs of the Ph.D. life by drinking beer of doing workout.*

I sincerely thank my partner **Marcello Stein**, *for his warm and constant support and motivation during the final "marathon", for reminding me about the many other important things in life and for cooking the most delicious pasta and clams far and wide.*

Finally, I thank my parents **Mioara** *and* **Karl Jochmann** *and my sister* **Andrea Jochmann**. *Without them, I would not have made it so far. I barely find words to express my gratitude. Thank you for your advices and your generous support during my life.*

I want morebooks!

Buy your books fast and straightforward online - at one of the world's fastest growing online book stores! Environmentally sound due to Print-on-Demand technologies.

Buy your books online at

www.get-morebooks.com

Kaufen Sie Ihre Bücher schnell und unkompliziert online – auf einer der am schnellsten wachsenden Buchhandelsplattformen weltweit!
Dank Print-On-Demand umwelt- und ressourcenschonend produziert.

Bücher schneller online kaufen

www.morebooks.de

OmniScriptum Marketing DEU GmbH
Heinrich-Böcking-Str. 6-8
D - 66121 Saarbrücken
Telefax: +49 681 93 81 567-9

info@omniscriptum.com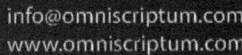
www.omniscriptum.com

Printed by Books on Demand GmbH, Norderstedt / Germany